Environment and Economy

As environmental issues move to the centre of the political debate, more attention is being focused on the role our economy has played in creating the ecological crisis, and what a sustainable economy might look like. In spite of the success of the environmental movement in drawing attention to the crisis facing us, there has been comparatively little attention focused on the way the operation of the global economy contributes to this crisis.

Environment and Economy provides a stimulating introductory insight into the history of thinking that has linked the economy and the environment. It begins by introducing readers to the pioneers of this field, such as Fritz Schumacher and Paul Ehrlich, who first drew attention to the disastrous consequences for our environment of our ever-expanding economy. Part II of the book describes the main academic responses to the need to resolve the tension between economy and environment: environmental economics, ecological economics, green economics and anti-capitalist economics. Part III is structured around key themes including an introduction to economic instruments such as taxes and regulation; pollution and resource depletion; growth; globalization vs. localization; and climate change. Each key issue is approached from a range of different perspectives, and effective policies are presented in detail.

Written in an accessible style, this introductory text offers students an engaging account of the way that the various traditions of economic thought have approached the environment, bringing them together for the first time in one volume. The text is complemented by boxes, case studies and recommended reading for each theme addressed. It will be of value to students interested in environmental sciences, geography, green issues and economics.

Molly Scott Cato is a Reader in Green Economics at Cardiff School of Management and Director of Cardiff Institute for Co-operative Studies.

Routledge Introductions to Environment Series
Published and Forthcoming Titles

Environment and Economy

Molly Scott Cato

Routledge
Taylor & Francis Group

LONDON AND NEW YORK

First published 2011
by Routledge
2 Park Square, Milton Park, Abingdon, Oxon, OX14 4RN

Simultaneously published in the USA and Canada
by Routledge
711 Third Avenue, New York, NY 10017

Routledge is an imprint of the Taylor & Francis Group, an informa business

© 2011 Molly Scott Cato

The right of Molly Scott Cato to be identified as author of this work has
been asserted by her in accordance with the Copyright, Designs and
Patent Act 1988.

Typeset in Times New Roman by
Keystroke, Station Road, Codsall, Wolverhampton

British Library Cataloguing in Publication Data
A catalogue record for this book is available from the British Library

Library of Congress Cataloguing in Publication Data
A catalog record for this title has been requested

ISBN: 978–0–415–47740–6 (hbk)
ISBN: 978–0–415–47741–3 (pbk)
ISBN: 978–0–203–83415–2 (ebk)

To Richard Douthwaite and Andy Henley
who between them taught me most of what I know about economics

'There is no wealth but life'

John Ruskin

Contents

Figures

Tables

Boxes

Case studies

Abbreviations

BAU	business as usual
CBA	cost–benefit analysis
CI	citizen's income
EKC	environmental Kuznets curve
EPA	Environment Protection Agency (US)
ETS	European [Emissions] Trading Scheme
GDP	gross domestic product
GHG	greenhouse gas(es)
GM	genetically modified
GNP	gross national product
HPI	Happy Planet Index
IPCC	Intergovernmental Panel on Climate Change
ISEW	Index of Sustainable Economic Welfare
nef	new economics foundation
SSE	steady-state economy
TAC	total allowable catch
TEQ	tradable emissions quota
TEV	total economic value
UNCED	UN Conference on Environment and Development
UNEP	United Nations Environment Programme
WTO	World Trade Organisation
WTP	willingness to pay

 # **Preface**

Nothing is more important in the contemporary world than finding a solution to the complex and often conflictual relationship between the operation of the economy and the health of the planet. For that reason, I would like this book to feature in the curriculum of every student of economics in the English-speaking world (and I don't say that with an eye on the royalties). My experience of teaching economics in a British university has led to an increased sense of concern that this subject, while of growing significance in most other disciplines, is still marginal within economics, where perhaps its understanding is more important than in any other school. I have written this book without assuming any prior knowledge of economic theory or its mathematical methods, so that it can be read and understood by students of environmental science, earth sciences and political theory. However, I would also like it to achieve the maximum readership possible within economics courses, whose students I would invite to welcome its non-mathematical treatment as a refreshing change.

This is an ambitious book, seeking to summarize the contributions towards solving the environmental crisis from a range of economic perspectives. It is therefore inherently **heterodox** in its approach. For those not familiar with the arcane ways of the economics discipline, this may require some explanation. The present state of academic economics is one of neoclassical orthodoxy, where career advancement depends on subscribing to a certain perception of how the discipline should work and which methods it should use. Many who focus on the environment primarily, and have turned their attention to economics as a mark of that concern, have been critical of this uniform approach, which they hold to some extent responsible for the pickle we are in. I therefore felt the need here to represent a diversity of approaches.

The preceding paragraph probably makes clear that I am sceptical about the contribution that neoclassical economics can make to protecting the planet. Rather than present myself as a neutral observer, I should declare my own adherence to the green economics school. I have made my best efforts in the chapters that follow to give each school a balanced account and a fair wind, and I apologize for instances when my own prejudices have unwittingly crept in. If you feel particularly inspired by one of the approaches, or consider that I have not given it adequate account or merit, then the further reading sections should provide ample sources to explore in more depth.

I have divided the book into two main parts – one based around theories and the other focusing on issues – with the intention that lecturers who use it as a core text can follow this division in their teaching. The first part provides an introduction to the subject and a chapter giving credit to the whistle-blowers who first identified the economy as the source of the environmental crisis. Part II offers thumbnail sketches of the different economic schools and their take on the economy–environment relationship. Part III addresses a range of issues of environmental concern and outlines contributions that a range of different economic approaches might offer by way of solution.

Molly Scott Cato
Stroud, June 2010

Acknowledgements

My heterodox economist friends – Ioana Negru, Alan Freeman and Clive Spash – have helped me believe that a better way of teaching and researching economics is possible, while my green economist friends – John Barry, Nadia Johanisova, Chris Hart, Richard Douthwaite and Mary Mellor – have helped me to dream our own utopian economy into existence. Thanks also to my friends who subscribe to a more traditional way of undertaking economic analysis, particularly Piers Thompson, who made useful comments on Chapter 3, and Richard Godfrey, who has helped me think through my ideas about economic growth.

Thanks to Dan O'Neill and all those campaigning for a steady-state economy, and to Kate Picket and Richard Wilkinson, for their boundless enthusiasm. Steve Harris is an endless source of encouragement and wisdom. Jen Parker and Lucy Ford have shared support and friendship with me, while Rebecca Boden has taught me what a career is for. Barbara Panvel is a constant source of inspiration as well as information. For spiritual guidance and support, my thanks go to Mavis Salmon, Helen Hastings-Spital and Mike Munro Turner.

It would be impossible for me to think with such freedom and recklessness if I were not surrounded by the fertile intellectual environment of Stroud. Thanks to Fred Pitel, for thoughtfully bespoke computer solutions, and to Nick and Rich for helping me find my way cheaply and efficiently through the cyberworld. Thanks to Imogen Shaw for her artistic skills, patience and tact. Thanks to Mum and Dad, Jane and Sue, Ralph and Josh, Rosa, Doreen, Raymond and Annie, for personal support and friendship.

Andrew Mould has been a supportive and assiduous publisher, ably supported by Faye Leerink, Emily Senior, Dan Benton and Penny Rogers. I would especially like to note my appreciation of series editor David

Pepper, who went through the text with a fine-tooth comb, and to the anonymous reviewers who pointed up parts of the typescript that required further attention.

Finally I would like to pay a special tribute to all my friends and colleagues who are able to face the environmental crisis with unflinching honesty and still find the courage to take action.

Part I
Setting the scene

1 Introduction: an economy within the environment

1.1. Environment and economy: friends or foes?

The purpose of this book is to introduce you to the way that economists think, and particularly how they think about the environment. It was US President Harry Truman who frustratedly called for a one-handed economist, because he was so tired of being told, 'On the one hand this, but on the other hand that'. But, for my money, this pluralism in economic debate is a healthy sign, which should be encouraged. Economics is a social science and therefore is inevitably going to revolve around essentially contested concepts that generate an endless debate. In fact, if anything we have seen too little plurality in recent years, with a near-hegemonic domination by the neoclassical approach. So, while this book may not be one-handed, I have certainly tried to be even-handed in explaining the perspective of a range of different types of economics.

Many books that explore the relationship between the economy and the environment begin by outlining and explaining the essential tension between the two. I am going to diverge from this path for the simple reason that I do not believe that such a tension is inevitable. More than 150 years ago, John Ruskin wrote the very useful small statement that sums up my view of the relationship between the economy and the environment: 'There is no wealth but life.' If we have come to a state of difficulty with our planet because we are seeking to consume more than she can provide, then the solution is straightforward: we need to rethink the way our economy works. The tension appears to be inevitable only if you accept that the structure and essential dynamics of our economic system cannot be changed. If our species is to have a future on this beautiful planet, we need to sort out the structural problem – and fast.

That is why I am a green economist. I make that point at the outset because the rest of this book will be taken up with a range of approaches to the relationship between economy and environment, or the economy–environment tension, as many of those represented will describe it. This is my book, and so I state my position at the outset. However, I will endeavour to give those with whom I disagree a fair hearing, and will refer you to other sources where you can find out more about them. Then it is your decision, and it is one you must take urgently – the need for our economy to establish a more comfortable relationship with our environment is pressing. My generation has only exacerbated this problem; the current generation of young people will spend their lives trying to solve it: I wish them every success.

As economists, what do we mean by 'the environment'? It is clearly the source of all the resources that are used in the production of goods that are sold in the marketplace and that we use to support our lifestyles. In the chapters that follow, we will see how the different schools of economic thought have considered (in some cases only barely) the impact of economic processes on the environment. For some, the 'exploitation' of resources is an acceptable approach to economic life, while, at the other end of the spectrum, there are economists who believe that nature has a sacred value that can never be reduced to a price and bought or sold in the marketplace.

1.2. Complementarities and tensions within the economy–environment relationship

While it is my view that the tension between economy and environment is not inevitable, I am equally sure that the source of the current environmental stress – the evidence for which accumulates daily – is in economic activity. Table 1.1 lists some examples of the way economic processes impact on the environment. The first in the list is the problem of climate change – by far the most serious environmental problem the human race is facing – the primary cause of which is the burning of fossil fuels to produce energy for heating and production of goods, and to facilitate our transport systems. Fossil-fuel burning is also the primary cause of acidification, the culprits in this case being the oxides of sulphur and nitrogen. In addition to these pollution problems, our exploitation of more of the earth's resources is destroying ecosystems, causing species extinction and shortage of land on which to grow crops. We are also rapidly depleting the **non-renewable** resources of the earth.

Ekins identifies the 'agents' of these 'symptoms of **unsustainability**', but he does not identify economic causes; but in most cases the problem has

Table 1.1 *Symptoms of environmental unsustainability and their causes*

Problem	Principal agents
Pollution	
Greenhouse effect/climate change (global)	Emissions of CO_2, N_2O, CH_4, CFCs (and HFCs), O_3 (low level), PFCs, SF_6
Ozone depletion (global)	Emission of CFCs
Acidification (continental)	Emission of SO_2, NO_x, NH_3, O_3 (low level)
Toxic contamination (continental)	SO_2, NO_x, O_3, particulates, heavy metals, hydrocarbons, carbon monoxide, agrochemicals, organo-chlorides, eutrophiers, radiation, noise
Renewable resource depletion	
Species extinction (global)	Land-use changes (e.g. development, deforestation), population pressure, unsustainable harvest (e.g. overgrazing, poaching), climate change, ozone depletion (in future)
Deforestation (global, regional)	Land-use changes, population pressure, unsustainable harvest (e.g. hardwoods), climate change (in future)
Land degradation/loss of soil fertility ((bio)regional, national)	Population pressure, unsustainable agriculture, urbanization, unsustainable development, climate change (in future)
Fishery destruction (regional, national)	Overfishing, destructive technologies, pollution, habitat destruction
Water depletion ((bio)regional, national)	Unsustainable use, climate change (in future)
Landscape loss	Land use changes (e.g. development), changes in agriculture, population pressure
Non-renewable resource depletion	
Depletion of various resources	Extraction and use of fossil fuels, minerals
Other environmental problems	
Congestion (national)	Waste disposal, traffic

Source: Ekins (2000)

its origin in some productive system or other. In Table 1.2, I have suggested possible economic systems and processes that might have generated the symptoms identified by Ekins.

Table 1.2 also demonstrates how complex a task we have in pinning down exactly which economic process has caused the problem we may be concerned about. This is not only because the economy is globalized and made up of a vast number of players; it is also because many of the causes of the environmental stresses are interrelated. For example, China seeks to develop through massively expanding its productive capacity and exporting goods to the West. This requires energy, so many new power

Table 1.2 Examples of economic processes that give rise to the problems identified in Table 1.1

Problem	Economic cause
Pollution	
Greenhouse effect/climate change (global)	Massive expansion of productive processes that emit greenhouse gases; increasing use of fossil fuels
Ozone depletion (global)	Emission of CFCs by manufacturers of aerosols and refrigerants
Acidification (continental)	Emissions from fossil-fuel electricity-generating plants; rapid expansion of personalized transport
Toxic contamination (continental)	A huge range of industrial productive processes
Renewable resource depletion	
Species extinction (global)	Loss of habitat caused by displacement of subsistence farmers; pressure on land caused by population increase
Deforestation (global, regional)	Corporate pressure to use previously forested land for cattle grazing or biofuel production
Land degradation/loss of soil fertility ((bio)regional, national)	Loss of traditional systems of agriculture due to population displacement and movement away from subsistence agriculture
Fishery destruction (regional, national)	Excessively intensive fishing and industrial pollution
Water depletion ((bio)regional, national)	Expansion of demand due to changing lifestyle and expanded industrial production
Landscape loss	Population pressure and removal of subsistence farmers from land
Non-renewable resource depletion	
Depletion of various resources	Needed as inputs to productive processes
Other environmental problems	
Congestion (national)	Excessive material consumption

stations are built, creating a huge increase in China's emissions of greenhouse gas emissions. To counter this, China also invests in a programme of hydroelectric energy generation, including the building of vast dam projects (using energy and producing CO_2 via the production of concrete). The dam projects displace hundreds of thousands of subsistence farmers, who are now drawn into an economic production and distribution system that itself relies on fossil fuels, where their previous existence did not. The dam projects also disturb the local ecosystem, causing species loss and loss of topsoil. Like truth, tracing the source of the economic causes of environmental problems is a process that is rarely pure, and never simple.

Does it have to be like this? I am sure that it does not; we need to reverse our thinking about the environment. Rather than seeing it as a 'problem' we need to realize that the environment is our home, and the source of everything that we value. Once we adopt that perspective, we can begin to reorient our economy so that it finds the planet a comfortable home, rather than a treasure chest of resources to be raided and a bottomless pit for our wastes.

1.3. Economics and environment: some useful concepts

Three concepts are useful in exploring the relationship between the economy and the environment: efficiency, optimality and sustainability. The first two of these concepts arise from conventional economic approaches, but have a particular twist when we introduce the environment into the equation, and the third is a word that has taken on a special salience – and become the source of heated argument – as the environmental crisis has grown in scope and urgency.

Conventional economists have their own particular way of using the idea of 'efficiency', and this is largely in terms of missed opportunities. If resources are being used 'inefficiently' then opportunities are being squandered. Greater efficiency could bring net benefits in terms of greater consumption. However, from the perspective of environmental economics, we might wish to extend our consideration of efficiency. For example, for an electricity-generating company, the cheapest fuel is the most efficient fuel – conventional economics will often use the concept of efficiency in a purely financial way like this. However, if the generation of electricity using this fuel creates environmental damage, for example by producing emissions that turn into acid rain, then it might not be efficient from the perspective of society.

The concept of optimality starts out from the position we reached with our definition of efficiency, but then moves on to question the nature of the allocation of any particular good or service, and whether it is socially ideal. In the case of power generation, an efficient solution to the problem of how we allocate the good or service might be to build a polluting power station in areas where few people would experience the pollution. But would this be socially optimal? We would need to have some sense of the social damage and introduce concepts of justice and fairness into our consideration. Conventional economics considers that we have an **optimal** situation if no person can have their situation improved without any other person's situation becoming less beneficial. But some environmental commentators might also consider that we should take into account other species, or even assume that the environment itself has a value that should not be diminished to achieve optimal human outcomes. And we might need to raise questions about the initial allocation.

Since the concept first passed into general usage, the meanings of the word **sustainability** have flourished like leaves on a tree. For the purposes of economics, the important thing when we think about sustainability is not only longevity, in the sense that an ecosystem or the planet as a whole will endure, but also that the quality of the environment has not been degraded by our activities. Amongst technically minded economists, the main argument concerning sustainability has been between what we might broadly refer to as neoclassical and ecological economists, and has centred around an argument over the extent to which nature is sacrosanct. Both groups are happy to consider the benefits we gain from nature as a form of 'capital', but they disagree about the extent to which other types of capital can be used as substitutes for natural capital. For example, can we afford to lose the rainforest so long as we can find a technological process that will be able to absorb CO_2 to the same extent (and provide all the other amazing resources and benefits the rainforest offers)? A neoclassical economist would argue that natural and man-made capital can stand in the place of one another, so that if we lose one we can rely on the other; ecological economists argue that they are both necessary, we need both of them to work together in our economy, and the loss of one cannot be made up for by more of the other. For an ecological economist, natural capital is of primary importance and is essentially limited. Beyond this rather technical discussion, the green economists would take the position more akin to that of the Iroquois, who thought that every decision should be made in the consideration of its impact on the seven following generations. From this perspective, sustainability means leaving the planet and its resources for our descendants in at least as good a state as we have enjoyed.

The discussion around the three key concepts that link the economy and the environment has made clear the distinct approaches that will form the structure of the remaining chapters of this book. We begin with neoclassical economics, which does not feel any particular need to address the environment as a special issue, because the market will naturally resolve all problems, including environmental problems. Beyond this heady optimism, we find the environmental economists, who introduce the environment as a source of special concern in their work, but continue to use the methods and tools of neoclassical economics when producing their analyses and prescriptions. Their approach can be contrasted with that of the ecological economists, whose origin lies in an attempt to unite the understandings of ecology and economics. Ecological economists take a much more sceptical attitude to conventional market-based approaches, and question the assumption made by environmental economists that the tension between economy and environment can be resolved by market solutions.

All these approaches find some place within university curricula, but recently more radical views have begun to be heard, which are sceptical about the ability of any academic study to solve the problems generated by the environment–economy tension. On the one hand, these radical approaches argue that the system of thinking prevents progress being made in analysis and prescription; on the other, they argue that fundamental change is also required in terms of the way that the economy is structured. These anti-capitalists and the green economists have been addressed separately in this book, but they share a critical view of capitalism as an economic system.

1.4. The environment in early thinking about economics

The tale of the encounter between economists and the environment goes back to the very earliest days of economic thinking, which is generally considered to have begun with the French physiocrats: 'The physiocrats of mid-eighteenth-century France – the first economic theorists – tried to explain economics in accordance with natural law and saw agriculture and Mother Earth as the source of all **net** value' (Daly and Townsend, 1993: 13). The 'physiocracy' envisaged by these aristocratic theorists literally meant the 'rule of nature', and they saw land as central to the economy. This is not as green as you might at first imagine, though, since their motivation was to oppose political intervention and, at this time before the economy had expanded so as to put pressure on the planet, they supported the free expansion of economic activity. In this sense, we might see them as having more in common with proponents of the free market, and their faith in the ability of the natural law of competition to guide an economy in the most benign direction.

The physiocrats represented the agricultural interest in France, and stood in opposition to the 'mercantilists', at a time when trade was beginning to expand rapidly. The mercantilists argued for a market regulated so as to ensure the most efficient and rapid production of manufactured goods, which could then be traded. They began to equate wealth with money, as opposed to the physiocratic view that all wealth came from the land. From a mercantilist perspective, land was less important than **capital**, and so one of the aims of trade was for the state to accumulate reserves of precious metals. The role of the state, according to the mercantilists, was to protect domestic production and trade through the use of tariffs and subsidies. At this stage, 'the land', which we might see as at least partly analogous to 'the environment', began to slip from its central place in thinking about the economy; apart from in the work of a few **heterodox** theorists it was not to regain that place until the advent of a green approach to economics in the late twentieth century. This loss of focus on the environment as the true source of wealth may help to explain the uncomfortable relationship that has developed between the economy and the environment.

Mercantilism was replaced from the late eighteenth century onwards by the theories of the classical economists (Adam Smith, David Ricardo and John Stuart Mill), who foregrounded the important role of labour for the first time in economy theory: 'The classical economists, witnesses to the problems of mercantilism as well as the beginnings of the Industrial Revolution, saw labor as the source of wealth and the division of labor and improvement in the "state of the arts" as the source of productivity' (Daly and Townsend, 1993: 13). Adam Smith's famous account of the greater productivity that could be achieved when the various tasks involved in making pins were divided so that a person specialized in just one task was the original model for all subsequent arguments for a division of labour in manufacturing. The classical economist David Ricardo penned the theory of comparative advantage, which argues that a country will achieve the highest standard of welfare if it concentrates on producing whichever good it produces most efficiently, and trading for the rest. This theory is flawed in various ways, perhaps particularly because it does not take into account the environmental consequences of vastly expanded trade (see further discussion of this in Chapter 12).

Thus we can see that the originators of the discipline of economics set the real stuff of the productive economy – people and their work, the land and its resources – at the heart of their study. However, the history of economics over the past century and a half has seen a movement away from these concerns. The first extant school of economic thought that is represented in this book is neoclassical economics, which is also the dominant paradigm in the Western academy in these early years of the twenty-first century. It is a study of economics that has drifted away from the real world – the environment within which all economic activity takes place and which provides all economic resources – focusing rather on modelling the world mathematically and interpreting human motivations and expectations.

I would identify myself as a **heterodox** economist, and my commitment to a pluralist approach to the teaching of economics is the reason that this book represents the views of economists from a range of traditions. This is unusual for an economics textbook, since today the neoclassical paradigm is dominant in our universities, and course materials and textbooks reflect this domination. Figure 1.1 gives an indication of how

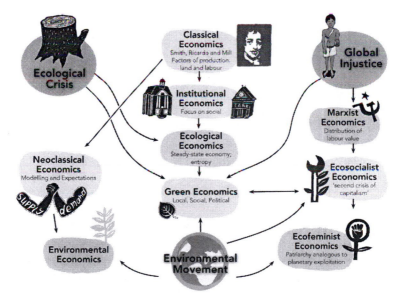

Figure 1.1 *Economic paradigms and the environment*

Source: Author's graphic drawn by Imogen Shaw

the schools of thought that are represented in the following chapters have grown out of the work of the original economic theorists.

1.5. A tale of many traditions

The neoclassical approach to the study of economics that dominates the twenty-first-century academy favours the market as the mechanism to determine how resources are allocated. For most neoclassical economists, the impacts of production processes on the environment can be defined as **externalities**, because firms do not consider them when they set the prices for the products, the process of which is the main subject of analysis in microeconomics, one of the two main branches of economic study. (Briefly, **microeconomics** concerns the study of what happens at the level of the firm; **macroeconomics** concerns what happens at the national and international level.) They fall outside this decision-making system, and so are considered 'external' (this is discussed more fully in Chapter 3). Neoclassical economists who take the environment more seriously fall under the heading 'environmental economists', some of whom also engage in 'natural resource economics', which means that they are particularly concerned with the depletion of resources. Their contribution to the debate is covered in Chapter 4. Although they make the environment the centre of their research agenda, environmental economists are still operating within the same intellectual paradigm, and using the same tools as neoclassical economists. For this reason they have been criticized by the 'ecological economists', whose work forms the basis of the discussion in Chapter 5. Ecological economics can be viewed as a marriage between two academic disciplines: that of ecology and that of economics. It makes the ecosystem we live within its primary point of reference, and then uses a scientific and mathematical approach to study how human resource needs could be met within this system in the least destructive way.

Chapters 3–5 cover the various ways in which the environment has been considered by academic economists. Chapters 6 and 7 implicitly suggest that these approaches have been rather limited, since they are contributions to economic discussion that have been made largely outside the academic world – by campaigners and concerned citizens. They also constitute a revival of an approach taken by the earliest economists – those who are now referred to as the 'classical economists' – who defined themselves in terms of 'political economy', recognizing that economics was not – and cannot be – a value-neutral, scientific study, but is closely bound up with issues of power. Chapter 6 covers the ideas of green economists, some of whom are writing and publishing in journals, but many more of whom work as campaigners in environmental organizations or as politicians. Chapter 7 covers contributions from the environmentalist wing of the anti-capitalist movement.

Much of the disagreement between the economists represented in the different chapters of this book concerns their view of the relationship between the economy and the environment, but they also disagree in their consideration of how social systems should interact with both economic and environmental systems. Figure 1.2 shows these three circles as interrelated, but separate, so that there are some parts of the economy that have nothing to do with society or the environment. This is the sort of perspective that might represent the approach to economic life known as 'sustainable development'. We recognize that these areas of life impinge on each other, and that sustainability is represented only by the place where they overlap. This diagram helps to explain the different strands of thinking that make up the chapters that follow. We can think of the neoclassical economists as focusing predominantly on the right-hand circle – their ideological approach suggests that the economy is primary, and will resolve social and environmental problems if the market is functioning efficiently. The environmental economists, whose work is covered in Chapter 4, share this picture, but put more emphasis on the

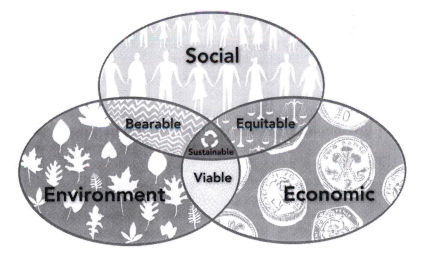

Figure 1.2 *Relationship between society, the economy and the environment*

Source: Drawn by Imogen Shaw

environmental circle – perhaps we can imagine it growing slightly in size as we adopt their perspective.

By the time we reach Chapter 5 – devoted to the work of the ecological economists – the focus has moved to the left-hand circle, with a primary focus on the environment and the need for the economy to fit within it without exceeding its limits, while also respecting social constraints and the need for politicians and policy-makers to constrain how economic systems function. For the green economists, whose work is covered in Chapter 6, the economy is a system of relationships, which should be embedded within a social and political framework, and the whole society must fit within the ecosystem. Finally, Chapter 7 outlines the views of economists we might broadly refer to as anti-capitalist, but who have a particularly pro-environmental bias; their focus is therefore largely on the upper and left-hand circles, their concern being predominantly with social justice as well as environmental protection.

Having explored the different theoretical approaches to the relationships between the economy and the environment, the second part of the book highlights a number of key issues, and explores how the different approaches suggest policies to deal with these in practice. To introduce this part, Chapter 8 discusses the different policy options that are available, and how we might assess which is most relevant and practical for each type of problem. Chapter 9 then deals with the most serious problem and the source of greatest contention amongst economists: the issue of whether the economy can continue to grow within a limited environment. Chapters 10 and 11 deal with the twin symptoms of environmental pressure caused by economic growth: the exhaustion of non-renewable resources is explored in Chapter 10, while Chapter 11 addresses the question of the waste products of economic activity. Chapter 12 explores an issue that has a strong political economy dimension: that of how the global economy is organized and whether globalization might need to be replaced by localization if we are to have a sustainable future. Chapter 13 tackles climate change, one of the biggest examples of the tension between an energy-driven economy and a limited environment. Chapter 14 focuses on the question of ownership and distribution of resources, and whether, from an environmental perspective, markets are the most efficient means of sharing the planet's wealth. Finally, in Chapter 15, I offer some closing reflections.

Summary questions

- What is meant by 'sustainability', and what is its relevance in an economic context?
- Which of the 'symptoms of unsustainability' identified by Ekins is it most urgent for us to address?
- Why was land of less concern to the economists known as 'mercantilists'?

Discussion questions

- Is the tension between economy and environment essential?
- Can we study economics without considering political influences?
- Why do you think neoclassical economics dominates in the universities?

2 The whistle-blowers

In the past few years, the environment has moved to the centre-stage of political debate, and climate change in particular – the most urgent but still only one of a plethora of environmental crises we are facing – has been identified by politicians as the most significant threat facing humanity, greater even than terrorism. It is important to remember, however, that this was not always the case. A mere 30 years ago, those who considered that the way we lead our lives might be causing environmental problems were regarded as fringe thinkers and dismissed as hippies or killjoys. Those who identified the central cause of the environmental problem as the interaction between economic activity and the environment were even rarer. This chapter pays tribute to a few of those thinkers – in different fields of work – who were the first to alert humanity to the seriousness of the problems we were facing.

2.1. Kenneth Boulding and spaceship earth

Kenneth Boulding began his life as a fairly conventional academic economist, teaching at Michigan and Boulder universities in the USA. His later work, an attempt at cross-fertilization between biology and economics, can be seen as a precursor to the development of ecological economics. Boulding considered that the discipline of economics chose its subject-matter too narrowly, and was ignoring the important environmental impacts of the economic system. He was an early proponent of the call to move towards a non-growth or 'steady-state' economy (see more on this in Chapter 9), and is famous for his statement, 'Anyone who believes that exponential growth can go on forever in a finite world is either a madman or an economist.'

Boulding's paper on the economics of 'spaceship earth' was published in 1966, and described the prevailing image that man has of himself and his environment. In it, the typical perception of the natural environment is that of a virtually limitless plain, on which a frontier exists that can be pushed back indefinitely. This 'cowboy' economy is an open system, involved in interchanges with the world outside; it can draw upon inputs from the outside environment and send outputs (in the form of waste) to the outside. In this perspective, no limits exist on the capacity of the economy to supply or receive material or energy flows. Such an economy measures success in terms of flows of materials, which it seeks to maximize. Stocks of materials are not conserved, since all resources are available in unlimited abundance relative to the human need for them. By contrast, Boulding identified that the planet is in reality a closed system: the only significant input to the economy of the earth is the energy from the sun; the resources that we have are fixed and exhaustible.

Boulding suggested as an alternative view that of the 'spaceman economy'. The earth should be viewed as a single spaceship, whose inputs are strictly limited and which must take account of its own wastes:

> Earth has become a single spaceship, without unlimited reservoirs of anything, either for extraction or for pollution, and in which, therefore, man must find his place in a cyclical ecological system which is capable of continuous reproduction of materials even though it cannot escape having inputs of energy.
>
> (Boulding, 1966)

Beyond the frontier of the spaceship itself, there exist neither reserves of resources nor waste sinks. The spaceship is a closed material system. In order to survive, from this perspective, humans must find their place in a perpetually reproducing ecological cycle. Boulding was writing at a time when there was no sense of the importance of the limits that the environment might impose on economic growth, although he detected some signs of change:

> The shadow of the future spaceship, indeed, is already falling over our spendthrift merriment. Oddly enough, it seems to be in pollution rather than exhaustion, that the problem is first becoming salient. Los Angeles has run out of air, Lake Erie has become a cesspool, the oceans are getting full of lead and DDT, and the atmosphere may become man's major problem in another generation, at the rate at which we are filling it up with junk.
>
> (Boulding, 1966)

Boulding was also critical of the straight-line thinking inherent in mainstream economics; this he described as 'a linear economy . . . which extracts fossil fuels and ores at one end and transforms them into commodities and ultimately into waste products which are spewed out the other end into pollutable reservoirs'. This way of organizing an economy was, he declared, 'inherently suicidal'. His alternative was a prototype for the spaceship earth, which he thought he had identified in the traditional village economy of Asia. Rather than a linear form this had a circularity built in – 'a high-level cyclical economy': materials should be used many times in different forms, rather than used just once, turned into waste, and disposed of. This was written nearly 40 years ago, and laid the groundwork for thinking about economic systems that has since been taken forward by ecological and green economists (see the further discussion of the closed-loop economy in Section 5.3).

2.2. The Club of Rome

In 1968, a group of concerned individuals came together to found the Club of Rome, an informal international association of academics and professionals with mixed expertise and backgrounds who shared anxieties about the course the world was taking. They launched the Project on the Predicament of Mankind, one outcome of which was the publication of the report *The Limits to Growth* in 1972. This report lays out in striking figures and compelling text the path of exponential growth that the human race is following in a number of different areas that may appear distinct but are in fact interrelated. As they suggest in the following parable, 'exponential increase is deceptive because it generates immense numbers very quickly':

> A French riddle for children illustrates another aspect of exponential growth–the apparent suddenness with which it approaches a fixed limit. Suppose you own a pond on which a water lily is growing. The lily plant doubles in size each day. If the lily were allowed to grow unchecked, it would completely cover the pond in 30 days, choking off the other forms of life in the water. For a long time the lily plant seems small, and so you decide not to worry about cutting it back until it covers half the pond. On what day will that be? On the twenty-ninth day, of course. You have one day to save your pond.
>
> (Meadows et al., 1972: 29)

This riddle is used to illustrate exponential growth, which is hard for people to conceive of. If we look at a whole range of indicators of resource use or natural systems, we see a familiar shape of curve, with an increase that then builds on itself, as in an exponential growth pattern.

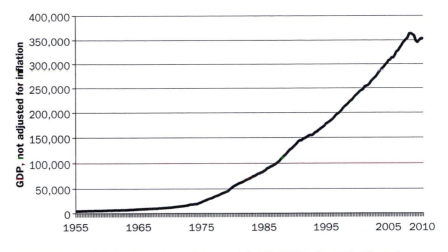

Figure 2.1 *Illustration of exponential growth in UK GDP in £m, not adjusted for inflation*

Source: Data from UK Office for National Statistics, Economic Accounts, Table A11, series ABNH, gross operating surplus, March 2010

Figure 2.1, which illustrates the increase in UK GDP, begins to approach this sort of pattern, and demonstrates what a shocking break with the expected pattern the 2008/9 recession was. Any graph illustrating population increase, consumption or the depletion of various resources follows a similar trend. Malthus's observation that human populations followed such a growth curve led him to draw pessimistic conclusions about the ability of those populations to feed themselves. Similar curves illustrating the depletion of natural resources caused concern to the Club of Rome members for two reasons: by the time the problems became obvious, there would be very little time left to solve them; and the pattern was apparent in so many resources and processes that it was clear that there was a systemic problem. They identified this problem as economic growth, which was causing all other systems to increase exponentially in denial of planetary limits (see more on the argument over the environmental consequences of economic growth in Chapter 9).

The report was, as would be expected, lambasted by the political and economic establishment, but it set the scene for the rise of environmental concern and its link with economic activity. The authors of the report produced an update in 2004, which included a useful summary of their argument:

> Our analysis did not foresee abrupt limits – absent one day, totally binding the next. In our scenarios the expansion of population and physical **capital** gradually forces humanity to divert more and more capital to cope with the problems arising from a combination of constraints. Eventually, so much capital is diverted to solving these problems that it becomes impossible to sustain further growth in industrial output. When industry declines, society can no longer sustain greater and greater output in the other economic sectors: food, services and other consumption. When those sectors quit growing, population growth also ceases.

In some ways, the warnings were premature, and allowed critics of the environmental movement to accuse environmentalists of over-interpreting their data and exaggerating the extent of the threat. The authors of the original *Limits to Growth* had provided an updated report in 1992, and concluded that, while we closer to 'overshoot and collapse', sustainability is still an achievable goal. With the benefit of hindsight, they found that their earlier report had focused on the depletion of raw materials as the limiting factor, relative to pollution, which has been shown to be the more serious issue. However, the role the original report played in alerting the world to the problems that rapid economic growth was causing to the environment cannot be exaggerated.

2.3. Ehrlich and *The Population Bomb*

Much of the early debate about environmental issues saw conflict between physical scientists (especially biologists) and economists. The economists tended to come off better in these intellectual tussles, although the onward march of environmental destruction suggests that they may not have been right. A classic example is the lengthy intellectual battle between Paul Ehrlich, a biologist, and the economist Julian Simon. Although a biologist by training, Paul Ehrlich was so concerned about the environmental crisis that he strayed into demography with his book *The Population Bomb*, published in 1968 and hugely influential on the burgeoning environmental movement. The book's message was stark, and reminiscent of Thomas Malthus in its apocalyptic vision: 'The battle to feed all of humanity is over. In the 1970s and 1980s hundreds of millions of people will starve to death in spite of any crash programs embarked upon now. At this late date nothing can prevent a substantial increase in the world death rate' (Ehrlich, 1986: ix). He was also, as this quotation indicates, shown to have been too simplistic in his assumptions about technological change and population growth, which greatly undermined the credibility of his critique. The reputation for scaremongering that was acquired by the early environmentalists would allow policy-makers to dismiss the later whistle-blowers, and their more scientifically grounded alarms about climate change in particular, as boys crying wolf.

Together with environmental scientists Barry Commoner and John Holdren, Ehrlich developed the IPAT equation in the early 1970s (for more detail, see Chertow, 2001). The equation is based on an unsophisticated thesis: that the environmental impact of economic activity is a combination of population size, consumption level and the efficiency of technological processes. A simple mathematical equation including these three variables generates IPAT:

$$I = P \times A \times T$$

$$\text{Impact} = \text{Population} \times \text{Affluence} \times \text{Technology}$$

To demonstrate the usefulness of the equation, we could consider car ownership in the USA, Brazil and China. According to figures from the World Bank, in 2007 there are 820 motor vehicles per 1,000 people in the USA, 198 in Brazil, but only 32 in China.[1] Assuming that the level of technology is the same in three countries in what is a globally competitive sector, we can calculate the IPAT values as shown in Table 2.1.

These figures for the relative impacts of the three countries in terms of car use indicate that, in spite of having the smallest population, the USA has by far the highest IPAT value. China, by contrast, has the world's largest population, which, although it has a significant environmental impact, is tempered by the much lower level of consumption of its citizens. Brazil, meanwhile, occupies a middle position, with a relatively high level of consumption and a population of two-thirds that of the USA, although its IPAT value is only a quarter the size. It is important to note that Ehrlich's

Table 2.1 The environmental impact of car ownership in Brazil, China and the USA

Country	Population (000s)	Car ownership	IPAT[a]
China	1,303,720	32	41.71904
USA	295,896	820	242.63472
Brazil	186,831	198	36.992538

[a]Divided by 1,000,000

Source: World Development indicators database; http://data.worldbank.org/indicator/IS.VEH.NVEH.P3

suggestion of this equation is extremely crude, and includes no scaling for environmental impact of the various factors. Its advantage is in pointing out that consumption levels should also be considered rather than just focusing on raw population rates. The Happy Planet Index (see Section 9.6) is a more sophisticated way of measuring how effectively different countries achieve their standards of living.

The response to Ehrlich's arguments came from the neoclassical economists led by Julian Simon. His book *The Ultimate Resource* (1981) argued what became known as a 'cornucopian' case (after the horn of plenty that was borne by Amalthea in Greek mythology). This proposed that the resources of the earth were bountiful and that human ingenuity could substitute for any shortage by producing inventive technologies. Simon's thesis is that historical trends suggest that technological progress will avert both the problem of resource scarcity and that of pollution; although problems of shortage and pollution will arise periodically, the nature of the world's physical conditions and the ingenuity of people allow such problems to be overcome, leaving us better off than if the problem had never arisen. Hence, an increasing population, the driving force behind technological development, is a blessing rather than a curse. To test Ehrlich's theory of the impact of population on resource scarcity, Ehrlich and Simon entered into a public wager on the assertion that a range of valuable minerals would dwindle over time: this is described in detail in Box 3.1 in the next chapter – but to spare you the suspense, I can reveal that Ehrlich lost.

While much of what Ehrlich argued was sensationalist and wrong, he and his colleagues were successful in alerting US policy-makers and world citizenry to the highly damaging impacts of an increasingly indulgent consumption pattern, combined with a rapidly growing human population. Ehrlich focused his attention on the population variable of his equation, founding the pressure group Zero Population Growth. The focus on the need to conserve resources was translated by such organizations as the Rocky Mountain Institute, the Factor 10 Club and the Industrial Ecology movement into a focus on improved design that would minimize resource and energy use.

As a coda to this story, a green or anti-capitalist economist might draw attention to the fact that, of the three variables that contribute to environmental impact in Ehrlich's original equation, technology is the one that is most amenable to variation by policy-makers with minimum social opposition. The alternatives – exercising serious control over population growth, or insisting that people curb their consumptive ambitions – are unpalatable to politicians who depend on votes for their continued power. Within a capitalist economy, it is also clear that developing a new technically advanced production process can generate profits for a business, while the business advantages in contraception or sales of hair-shirts are seriously limited.

2.4. E. F. Schumacher and *Small is Beautiful*

While E. F. (Fritz) Schumacher is best remembered for coining the adage of the green movement, 'small is beautiful' (a phrase that some claim was actually derived from his teacher Leopold Kohr), his contribution to economic theory is perhaps better represented by his statement that, 'It is inherent in the methodology of economics to ignore man's dependence on the natural world.' He was highly critical of the economics profession, and agreed with a view he attributed to Keynes – that economics should become a 'modest occupation similar to dentistry', rather than become the dominant subject of modern life. He felt that the economics profession, with its misguided and inadequate method of cost–benefit analysis and its attempt to put a price on what was priceless, was destroying the quality of modern life. He argued instead for an 'economics of permanence'.

Schumacher's thinking about the relationship between the environment and the economy grew out of his two decades of experience as a civil servant working for the British Coal Board: as an economist there, he was able to observe how the British economy used resources in a profligate and short-sighted way. He linked this to an economy addicted to growth, and was thus an early critic of economic growth, preferring to consider well-being as an indication of a successful economy:

> It is clear that the 'rich' are in the process of stripping the world of its once-for-all endowment of relatively cheap and simple fuels. It is their continuing economic growth which produces ever more exorbitant demands, with the result that the world's cheap and simple fuels could easily become dear and scarce long before the poor countries had acquired the wealth, education, industrial sophistication, and power of capital accumulation needed for the application of alternative fuels on any significant scale.
>
> (Schumacher, 1973: 28)

Schumacher was concerned with the ever-increasing scale of production systems, arguing that, 'The economics of giantism [that is, mass production and centralized, large-scale consumption] and automation is a leftover of nineteenth-century conditions and nineteenth-century thinking and it is totally incapable of solving any of the real problems of today' (ibid., 61). Rather, he believed that development should take place outside the cities and create an 'agro-industrial structure' of small towns based in the countryside. In this, his work prefigured many of the proposals made by today's green economists. He also shared the green economist's scepticism about technology, arguing not against technology itself, but against ever more sophisticated technology, almost for its own sake. His own conception was of 'intermediate' or 'appropriate' technology, which he also referred to as 'technology with a human face' (to some extent, this work prefigured the development of the industrial ecology movement: see Section 11.3).

Schumacher's critical approach to modern economic life was extensive. He despised the individualism that the complex, global market system both presupposed and encouraged, and he bemoaned the substitute of quantity for quality that a modern market economy brought in its wake:

> ... the reign of quantity celebrates its greatest triumphs in 'the Market.' Everything is equated with everything else. To equate things means to give them a price and thus to make them exchangeable. To the extent that economic thinking is based on the market, it takes the sacredness out of life, because there can be nothing sacred in something that has a price.
>
> (Schumacher, 1973: 61)

While Schumacher's original inspiration was derived from traditional economic concerns such as resource scarcity and efficiency, his later work dwelt more on the social and spiritual aspects of economic organization, as in his essay 'Buddhist Economics', where he compares the perspective of a neoclassical economist with that of a putative 'Buddhist economist'. Building on the concept 'right livelihood', which is one of the Noble Eightfold Paths of Buddhism, he questioned how an attitude to work informed by a Buddhist philosophy might be possible within an industrialized economy. What, he asked, would this imply in terms of how we organize our economy? How would work within such an economy 'give man a chance to utilise and develop his faculties; to enable him to overcome his ego-centredness by joining with other people in a common task; and to bring forth the goods and services needed for a becoming existence?' (ibid.)

2.5. Howard Odum: thinking in systems[2]

Howard T. Odum was a pioneering American scientist who made major contributions to ecology, energetics and systems theory, founding the fields of ecosystems ecology, ecological modelling and ecological engineering. Odum developed methods for tracking and measuring the flows of energy and nutrients through complex living systems; his work investigating the effects of nuclear radiation on such flows in coral reefs, rainforests and estuaries reflected his lifelong quest to find ways of understanding and dealing with man-made impacts on the environment. During the late 1950s, Odum became particularly interested in the interface between the economy and the environment, and began to look for ways of understanding the links between flows of money and goods in society and the flows of energy in ecosystems. This eventually led to his development of an 'Energy Systems Language' which he claimed allowed the accurate and precise mapping of all kinds of energetic and resource flows – including money – in interlinked human/natural systems.

Chapter 5 will explain how the application of insights from the natural sciences to social and economic structures is crucial to the development of ecological economics. Odum's work has been especially important in clarifying the extent to which social and economic systems are affected by, and must understand and respect, the laws that govern energy and resource flows in natural systems. Odum's Energy Systems Language was based on the insight that nearly all the energy available to fuel natural and human activity begins as heat and light from the sun; this solar energy must then be transformed by natural systems (e.g. photosynthesis by plants) and human technologies (e.g. solar photovoltaics), and stored in various forms (biomass, batteries) before it can be used. Odum proposed that a measurement of the amount of transformed solar energy embodied in any product of the biosphere or human society – for which he coined the term 'emergy' – could provide a kind of 'universal currency'; this would allow fair and accurate comparison of the human and natural contributions to any particular economic process. An example might be the comparison of fossil fuels, with biodiesel from crops, and biodiesel from recycled vegetable oil. An emergetic analysis could providing 'costings' for each of these, comparing the amount of energy required to produce them with the energy they provide in terms of heat or automotive power. This approach was so original that it has still not been fully incorporated into thinking about responses to climate change, where understanding the **embodied energy** in products is arguably more critical than only considering the direct energy flows in electricity generation or the work of an internal combustion engine. (Section 5.2 has more detail on the impact of entropy on economic systems and the importance of thinking in a systemic way.)

Odum was also one of the first commentators to clearly identify and measure the absolute dependence of all aspects of modern economies on the vast stores of solar energy embedded in non-renewable fossil fuels – his demonstration (in *Environment, Power and Society*, 1971) that 'industrial man . . . eats potatoes largely made of oil' galvanized a generation of scientists and activists. In his final book, *A Prosperous Way Down*, written in 2001, Odum set out his plans for 'energy descent' – that is, for the humane transition to a society based on the use of **renewable**, rather than fossil-fuel energy. His studies of the cycles of growth, storage, consumption and depletion in natural systems had convinced him that modern societies have reached the crest or climax of a period of massive growth driven by fossil fuel energy, and that downturn is now inevitable. By suggesting that the concept of emergy could help current and future generations answer the crucial question, 'What is real wealth?', Odum claimed to offer a quantitative and scientifically-based foundation for the transition to a low-energy, low-carbon economy.

Odum can be seen as a pioneer of attempts to price the value that ecosystems provide for us – a process now known as the costing of 'ecosystem goods and services'. Although focusing the attention of economic thinking on the value of natural systems has certainly moved the environmental agenda forward, some ecological and most green economists argue that 'costing the earth' in this way can be highly problematic, seeing it as inappropriate to attempt to put literally priceless services under the control of market forces. Odum's response to this argument was that if economic values were based on measures of the quantity and quality of embodied solar energy, rather than on the basis of monetary worth measured as willingness-to-pay, we would be much nearer to an understanding of what constitutes 'real wealth' – and much less inclined to recklessly squander the precious stores of that wealth available to us in, for instance, rapidly depleting fossil fuel deposits. Odum's approach has not only been influential on ecological economists such as Robert Costanza, but has also provided an important foundation for the grassroots eco-social design movement of Permaculture (discussed in Section 11.3).

2.6. Murray Bookchin: prophet of localization

Murray Bookchin might seem an unlikely candidate as an inspiration for an economic system that is environmentally benign, having his origins in the US Communist movement of the 1930s. However, his work was unusual in responding to the disillusion with Communism as it was practised through the 1930s by moving in a libertarian direction; then, in the 1960s, he drew attention to the negative environmental impacts of industrialized societies, and proposed small-scale anarchist communes as a socially and environmentally preferable system of social organization. Bookchin's philosophy combined his earlier Marxism with anarchist organizational structures: 'he drew on the best of both Marxism and anarchism to synthesize a coherent hybrid political philosophy of freedom and cooperation, one that drew on both intellectual rigor and cultural sensibility, analysis and reconstruction. He would call this synthesis social ecology' (Bookchin, 1997: 4).

Developing his ideas during the apotheosis of US technologically fuelled capitalism, Bookchin's extraordinary insight was to suggest that 'an ecological crisis lay on the horizon', that chemical food additives might be causing cancer and degenerative diseases, and that technological 'progress' might yet have harmful environmental consequences. He identified the profit-driven capitalist system as the cause of the ecological crisis, and suggested an anarchist and decentralized solution in his book, *Our Synthetic Environment*, published in 1962. Bookchin's manifesto was his 1964 essay 'Ecology and Revolutionary Thought', which first outlined what he would later call 'social ecology'. Its central thesis was that, 'The imbalances man has produced in the natural world are caused by the imbalances he has produced in the social world' (Bookchin, 1971: 62). Bookchin's prescience in a range of areas where ecological concern is commonplace today is striking: he warned of the greenhouse effect as early as 1964, when it was not widely discussed; suggested a link between radionuclide pollution and the increase in cancer; and predicted negative health impacts of stressful urban life. His solutions were equally radical and ahead of their time, as in his view of the renewably powered eco-communities (see Section 6.4) that many greens are striving to create nearly 50 years after he first proposed them:

> To maintain a large city requires immense quantities of coal and petroleum. By contrast, solar, wind and tidal energy can reach us mainly in small packets . . . To use solar, wind and tidal power effectively, the megalopolis must be decentralized. A new type of

> community, carefully tailored to the characteristics and resources of a
> region, must replace the sprawling urban belts that are emerging
> today.
>
> (Bookchin, 1971: 74–5)

The fact that Bookchin has not been widely recognized as a forerunner of
modern environmentalism may derive from his stalwart rejection of
'environmentalism' as a form of solving ecological problems without
addressing their fundamental social and economic causes. He thus came
down clearly against the 'ecological modernization' discourse, as
discussed later in Chapter 8, and, of course, before it was defined as such.

In terms of his prescription, Bookchin joins Schumacher in identifying
the scale of modern human communities as a root cause of socio-
economic and environmental problems. From an organizational point of
view, he opposed the excessive specialization of modern production and
the 'intensive division of labour of the factory system'. However, his
concern for the size of organization was also about the disempowering
nature of large structures, both industrial and political:

> A small or moderate-sized community using multipurpose machines
> could satisfy many of its limited industrial needs without being
> burdened with underused industrial facilities . . . The community's
> economy would be more compact and versatile, more rounded and
> self-contained, than anything we find in the communities of
> industrially advanced countries.
>
> (Bookchin, 1997: 25)

This discussion prefigures the debates about localization as a response to
the damaging effects of globalization (presented in Chapter 12), as his
view on political organization and structure anticipates the call for
'resilient local commnities' of many contemporary environmentalists:

> I do not claim that all of man's economic activities can be completely
> decentralized, but the majority can surely be scaled to human and
> communitarian dimensions. This much is certain: we can shift the
> center of economic power from national to local scale and from
> centralized bureaucratic forms to local, popular assemblies. This shift
> would be a revolutionary change of vast proportions, for it would
> create powerful economic foundations for the sovereignty and
> autonomy of the local community.
>
> (Bookchin, 1997: 25)

Bookchin's work was revolutionary not because of its analysis or
prescription, but because of its original synthesis of concerns for the
environmental and social consequences of the modern industrialized

economy, and its recognition that these form a nexus, which should be tackled **holistically** before we can build a truly sustainable economy.

2.7. Hazel Henderson attacks the snake-oil doctors

When Hazel Henderson titled a 1982 paper 'Three hundred years of snake oil: defrocking the economics priesthood', she was merely stating in florid terms the widely held view amongst heterodox economists – as opposed to the neoclassical brotherhood – that the discipline is more akin to a faith than a science. Her contribution to the development of an environmentally sensitive economics is to identify the political nature of the economics profession, and to blow apart the myth of economics as an objective science: 'The word is out that economics, never a science, has always been politics in disguise. I have explored how the economics profession grew to dominate public policy and trump so many other academic disciplines and values in our daily lives' (Henderson, 2006). Henderson's suggestion that economics is used as a cover for politics can be seen in the examples where phrases derived from economics, such as 'economies of scale', are used as explanations for why systems that might be more sustainable, such as locally based economies, cannot be considered legitimate policy objectives.

Henderson has always been critical of the 'free market system' (the quotation marks are hers) and quotes Karl Polanyi as identifying that the so-called laissez-faire market is in fact a system of social relationships. In her highly critical view, such market systems are symptomatic of what she calls 'flat-earth economics', meaning a way of looking at the economy that refuses to pay attention to mounting scientific evidence that it has ceased to be appropriate. Her view of a sustainable economy is one where the power is stripped away from global corporations, and reinstated again in enterprises owned and controlled by their own workers based in strong, mutually oriented communities: 'worker-owned self-managed enterprises, and of bartering, sharing, self-help and mutual aid' (Henderson, 1988: 101). Such an economy will be focused on provisioning based on a culture of relationship rather than on markets mediated through money (the relationship between ownership and sustainability is covered in depth in Chapter 14).

Henderson illustrates this through her image of the global economy as a cake (see Figure 2.2). Conventional economics considers only the upper layers of the cake, which is the market economy that operates on the basis of work and monetary exchange. But this layer is only superficial and

Total Productive System of an Industrial Society
(Layer Cake With Icing)

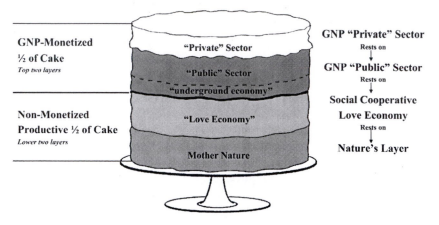

Figure 2.2 *Henderson's illustration of the global economy as an iced cake*

Source: Thanks to Hazel Henderson for permission to reproduce this graphic free of charge

relies on the uncounted work of women, producers in poorer nations, i.e. the reciprocal or community economy and its relationships of trust, and underneath that the planet itself; it is because we neglect these fundamental bases of our economies that our communities and our planet are threatened. Thus the message at the heart of Henderson's work is one that is increasingly relevant as the tension between a globalized capitalist economy and the planet becomes more apparent:

> The Solar Age signifies much more than a shift to solar and renewable resource-based societies operated with more sophisticated ecological sciences and biologically-compatible technologies. It entails a paradigm shift from the fragmented 'objective' **reductionist** knowledge and the mechanistic, industrial world-view to a comprehensive awareness of the interdependence of all life on earth.
>
> (Henderson, 1988: xxi).

This brief quotation includes many of the philosophical underpinnings of the environmental movement, which inform the more radical approaches to the economy–environment relationship. Their critique of the reductionism that is typical of neoclassical economics – reducing complex systems to simplistic equations or graphs – leads to the suggestion that we should rather seek a **holistic** approach, which reflects the complexity of the natural world.

This chapter has presented thumbnail sketches of the work of a few of the insightful and wise thinkers who first noticed that the global, industrialized economy was on a collision course with our environment. It is worth noting, I think, how long it has been since these alarms were first raised, and how long it has taken policy-makers to undertake anything like a serious response. This book focuses on economics and the different ways that economists have approached this agenda. I suppose that right here we should notice that the most significant aspect of the response has been its extremely slow pace. Over the past 60 years or so, economics has been divorced from 'political economy', and it seems to me (as it does to Hazel Henderson) that the absence of the political dimension might help to explain the snail's pace with which economics has responded to the energetic warnings of the whistle-blowers. We do not reach the work of economists whose work might be considered to be a revival of political economy until Chapters 6 and 7. Between now and then, we have three chapters outlining the work of more conventional academic economists.

Summary questions

- What did Boulding mean by a linear economy?
- How do you understand 'exponential economic growth'?
- Is Bookchin's vision of a green economy realistic?

Discussion questions

- Which of the whistle-blowers do you think was the most prescient, and why?
- How many of the whistle-blowers do you think could be considered economists?
- Why do you think it has taken so long for the economists to respond to the alarms raised by the whistle-blowers?

Part II

**Economic schools
and the environment**

3 Neoclassical economics

This chapter addresses the orthodox economics paradigm, and the reasons why its theoretical underpinnings might be especially challenged by the environmental problems that we are facing. We begin by identifying how its particular view of the world deals with environmental issues. Economics has its own particular jargon and method (involving a considerable amount of mathematics and graphs), and, although I have kept these to a minimum, this may prove challenging to some readers; I hope that the argument will still be clear. This chapter presents the bones of the neoclassical approach and the proposals from the economics mainstream in response to the environmental problem. Environmental economics, which is covered in the following chapter, shares many of the assumptions and methods of the neoclassical approach. To some extent, the division of this body of thought into two chapters is pragmatic, although it seems fair to say that the theories and theorists covered in Chapter 4 have made the environment central to their study, and that their work tends to have emerged since environmental problems came to the fore, from around the late 1960s onwards.

We begin with an outline of the market system, which is the ideal for economists in this tradition. Section 3.1 looks at the issue of resources from an orthodox economics perspective and describes the market view of how resources are distributed within an economy, including a discussion of the technique of cost–benefit analysis, while Section 3.2 looks at how this system deals with scarce resources and their allocation. Section 3.3 then explains the work of two early twentieth-century economists who addressed environmental questions in their work, and Section 3.4 then considers one aspect of pricing economic activity: the process of discounting. Finally, Section 3.5 presents a case study of a market solution to one specific environmental problem: sulphur dioxide emissions.

3.1. Markets, efficient allocation and assessing outcomes

Neoclassical theory was intended to be neutral and value-free: like Newton's laws of physical motion, it aims to define a set of laws governing economic activity. In this theory, forces of supply and demand interact to achieve optimal outcomes for all. Economic decisions, including those about non-price environmental goods and future generations, are made on the basis of 'utility maximization', a means of ensuring that the preferences of as many economic agents as possible are fulfilled to the maximum extent possible given the limitations of available resources. Although the economy is a dynamic system with a multitude of individual players, nonetheless it reaches an equilibrium where these forces are in balance. While the objective of the economic laws is to achieve a balance, this should not be equated with a fixed system – rather the economy is expected to grow relentlessly. This Promethean view of the boundless expansion of human ingenuity can be identified with the optimism of the nineteenth century, when technology was allowing increasing mastery over nature and man's power seemed limitless.

The most important aspect of a market system, and the reason given for its superiority to other forms of economic organization, is efficiency. So what do economists mean by efficiency? The understanding of efficiency was greatly influenced by the work in welfare economics of Vilfredo Pareto. In simple terms, they mean that the allocation of a resource is 'efficient' if there is no way that a person who is receiving some of that resource could receive more without some of that resource being taken away from somebody else. We begin with the assumption that there is a limited amount of this good to go around. This theory is illustrated in Figure 3.1, which considers the allocation of a scarce resource between just two people: person 1 (represented by the vertical or y-axis) and person 2 (represented by the horizontal or x-axis). The figure illustrates the 'efficiency frontier' for the allocation of a good between two people; when the goods are allocated efficiently, the total value allocated is £100. At point d, person 1 receives all of the good; at point f, person 2 receives all of the good. At points in between, and along the frontier, allocations are more or less equal, but all are efficient. Anywhere within the frontier, for instance point a, where each person receives an amount of the good worth £25, the allocation is inefficient, since either one person or the other could receive more of the good by moving to the frontier. From point a, a move could be made to either point b or point c, allowing an increase in utility represented by the shaded triangle. It is notable that this

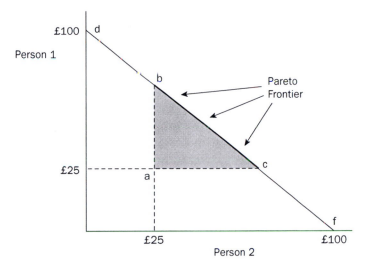

Figure 3.1 *Pareto-efficient distribution*

allocation model does not concern itself with the relative shares acquired by the two people, i.e. with the issue of allocative fairness.

For a neoclassical economist, it is a fundamental truth that markets produce the most efficient outcomes, in the Pareto sense; that is why market systems are held to be superior to other forms of economy. There is no need for policy-makers to act, since the 'invisible hand' of the market will automatically achieve this efficient outcome.

Almost all neoclassical economists accept the need for governments to play some role in the economy, and that these policy-makers require information before making a decision about the likely outcomes. When deciding which policy choice is better from a societal perspective conventional economists turn to the cost–benefit analysis (CBA). This is a technique to measure the **optimality** – in a Pareto sense – of the outcomes of a particular policy, such as building a new road or cutting down a forest. It involves summing all the costs and benefits of a project: so long as the benefits outweigh the costs, the decision should be made for the project to go ahead. The CBA begins by defining clearly what is being measured: what time period is being considered, exactly what changes will be made and whose welfare is being included in the equation. The next stage is to identify all the physical impacts of the project before the most difficult stage of all: costing the impacts. All the calculations are worked in monetary values, which means that a monetary cost must be calculated for any positive of negative impact of the proposed action. In addition, a discount rate is applied, to allow for the fact that costs and benefits may not have equal real value at different periods in time. This technique, which can have a huge impact on the likelihood of a policy being introduced, is discussed in the next section.

Environmental economists have criticized conventional CBA for not costing in certain features – e.g. the economic value of a scenic view – which are hard to price. Heinzerling and Ackerman (2002) call CBA a 'deeply flawed method that repeatedly leads to biased and misleading results'. They criticize the central assumption of the method, that the issues of concern when decisions are made can be translated into monetary terms. They challenge CBA's two central claims to superiority as a decision-making tool: the achievement of efficient allocations and objectivity/transparency. But Heinzerling and Ackerman claim that CBA cannot assure efficient allocations because it can only measure costs at the individual scale, and in monetary terms, which is not the way humans actually assess the consequences of decisions, either for themselves or their environment. The method appears to be objective and transparent, because citizens can explore what values are put on certain costs and benefits, and the numbers are available for checking. As far as transparency and objectivity are concerned, Heinzerling and Ackerman suggest that the method assumes a particular worldview (a monetized, individualist one), which is far from objective, and the process itself is so cumbersome and complex that only experts or wealthy citizens can afford to engage with it.

3.2. Price, scarcity and substitutability

Several of the key concepts that guide a neoclassical economist are problematic when applied to what is increasingly being recognized as a limited natural environment. An economist will talk about the 'exploitation of resources' without any concern about the moral implications of the phrase. As we are told by the *Oxford Dictionary of Economics*, 'This is an entirely value-free usage; it is contrasted with the pejorative usage'. From this perspective, the earth's resources are freely and unproblematically available for our use. Another key concept in economic theory is that of 'economies of scale', which is defined as, 'The factors which make it possible for larger organizations or countries to produce goods or services more cheaply than smaller ones'. Although attention is also paid to diseconomies of scale, orthodox economic theory suggests that that there is an in-built tendency for enterprises to grow in order to benefit from these scale economies and thereby increase profits. This runs counter to the concern of ecological and green economists for limits to growth.

Neoclassical economics tends to deal with environmental problems under the heading of 'market failure'. Since the assumption is that markets are perfect systems, if something is less than optimal then the market has failed, and the appropriate response is to remove whatever it is that is acting as a barrier to the free operation of the market. Market failure is dealt with in more detail in Section 4.1. In this chapter, we will limit ourselves to consideration of the most important example of market failure from the perspective of the environment: that of the **externality**. An externality is a consequence of economic activity that does not impinge on the person or business conducting that activity.

Economic theory suggests that there are positive and negative externalities: pollution is the classic example of a negative externality; positive externalities are harder to find, but perhaps we might think of the pleasant smell from a local bakery as falling into this category. From the perspective of the environment, it is the negative externalities that concern us. Any sort of pollution would be defined as an externality by a neoclassical economist – a factory can produce plastics and release dioxins into the air free of charge. Their release does not impinge on the factory's production, and does not feature in its cost calculations: for this reason, its quantity cannot be controlled by the cost–revenue calculation that determines how a business will operate within the market. The market mechanisms therefore cannot constrain the output of pollution, hence the market failure. The negative consequences of the emissions are borne by society at large, or, more likely, by a small number of people who live near the factory. The fact that there are levels of pollution that seriously impair the health of environmental and social systems must, within the economic system, be seen as an example of 'market failure' – according to a neoclassical economist, if the market were operating successfully then any negative effects would be self-adjusting.

Neoclassical economists are also relaxed and optimistic about resource scarcity: they believe that the market will allocate goods efficiently, and the price signal is a key part of this process. Prices communicate information about how scarce a good or service is, relative to the demand for it. In other words, the price is set by the relationship between the supply of a good or resource and the demand for it. Thus, the price mechanism can do the job of protecting scarce resources for us. As a resource becomes more scarce, its price will rise and demand will therefore fall – automatically protecting the resource.

Scarcity is also a relative rather than an absolute concept, since if one resource becomes depleted another can be used as a substitute:

> If it is very easy to substitute other factors for natural resources, then there is, in principle 'no problem'. The world can, in effect, get along without natural resources. Exhaustion is an event not a catastrophe . . . If, on the other hand, output per unit of resources is effectively bounded – cannot exceed some upper limit of productivity which is, in turn, not too far from where we are now – then catastrophe is unavoidable . . . Fortunately, what little evidence there is suggests that there is quite a lot of substitutability between exhaustible resources and renewable or reproducible resources.
>
> (Solow (1993: 74), quoted in Gowdy and Hubacek, 2000)[3]

Hence, as demonstrated by the story in Box 3.1, neoclassical economists are not concerned about the depletion of natural resources, since they believe human ingenuity will resolve this problem by finding substitutes (see the further discussion of this in Chapter 10).

Box 3.1

A cornucopian matches up to profit of doom: the Ehrlich–Simon wager

Neoclassical economists subscribe to an optimistic view, where the limiting factor on human progress is human imagination, which can overcome any biophysical or environmental constraint:

> The major constraint upon the human capacity to enjoy unlimited minerals, energy and other raw materials at acceptable price is knowledge. And the source of knowledge is the human mind. Ultimately, then, the key constraint is human imagination acting together with educated skills. This is why an increase in human beings, along with causing additional consumption of resources, constitutes a crucial addition to the stock of natural resources.
>
> (Simon, 1996: 408)

Simon was challenging directly the concerns of environmental campaigners that the economic system was itself putting pressure on the planet and running up against planetary limits. As the quotation makes clear, Simon was unconcerned with either population growth or limits on the supply of energy or resources. He famously debated these issues publicly with Paul Ehrlich, author of *The Population Bomb* and a prominent environmental writer and campaigner (see more in Section 2.3). The debate culminated in a public wager that the price of five metals commonly used in production processes would not increase over a fixed time period. Simon was relying on the economic theory that suggested scarcity would be reflected in price signals and that, therefore, if the supplies of these metals were declining then their prices would increase.

Ehrlich accepted the wager and chose the following metals: copper, chromium, nickel, tin and tungsten, and a date ten years ahead, 1990. At the end of the ten-year period, the price of all the metals had actually fallen; Ehrlich lost the bet and paid Simon. Simon had proved that the prices of the commodity metals were falling, but did this actually prove anything about their exhaustion? Economic theory would suggest that, as a material becomes more scarce, 'substitution' takes place, i.e. technologies and processes are designed that reduce the demand for this material. In the case of copper, for example, fibre optic cables were invented and developed, which actually reduced the demand for copper. There are several variables that influence the price of commodities – rather than scarcity alone – so the falling prices need have had no direct connection with the biophysical limits on the availability of these metals.

Ehrlich's mistake was to accept the wager on Simon's home ground – his agreement to gamble on the future value of commodities indicated his limited knowledge of how markets function, rather than a fundamental mistake about whether the economy needs to respond to environmental limits. According to Lawn (2010) both Ehrlich and Simon were operating with an overly simplistic view of resource price-setting. As a simple example, much of the fluctuation in the price of commodities is the result of futures trading rather than a response to the level of supply.

3.3. The environment begins to impinge on the economics profession

Chapter 2 included a discussion of the first economists who considered the impact of the economy on the environment. In this section we consider the contribution made by two twentieth-century economists: Pigou and Coase.

Arthur Pigou was one of the first economists to discuss the problem of pollution. His work on welfare economics, published in 1920, argued an early version of the **polluter-pays principle**, and proposed the introduction of taxes on businesses that generated pollution. This was justified on the basis that, in producing the pollution, factories transfer a cost from themselves to the public, since they would have had to pay to dispose of the waste if they had not emitted it into the environment.

Because they are spared this expense, the government is therefore justified in reassigning that cost back to the factory in the form of a tax. Although neoclassical economists are generally opposed to government intervention in markets, it can be justified in this case because the pollution represents an **externality** and that is a legitimate reason for government involvement.

The key concern from this perspective is to work with markets to increase their efficiency, and so the objective is to set the right level of the tax to achieve the **optimal** level of pollution that allows the product to be made (for a discussion of why the optimal level of pollution cannot be none at all, according to this theory, see the discussion in Section 11.1). The definition of a Pigouvian fee is 'a fee paid by the polluter per unit of pollution exactly equal to the aggregate **marginal** damage caused by the pollution when evaluated at the efficient level of pollution' (Kolstad, 2000: 118). For a neoclassical economist, all decisions are made at the margin. The marginal product, or employee or unit of pollution, is the final straw, at the point where a decision to work, invest or pollute becomes financially advantageous. To operate efficiently in a neoclassical market is to identify where this marginal point is, and to make your decision to be on the right side of this line.

Figure 3.2 illustrates how the level of such a fee is established. In the graph, the lower line is the cost curve of the producer, which relates the quantity of a product that s/he produces to the cost of that production. The fact that it slopes upwards from left to right indicates that, at high levels of production, as you produce more of a good, the average cost of each unit of production increases (because of the complexity of running a large and complex output unit). The steeper curve is the cost to society from the pollution generated by the production process (the marginal social cost), which is at present an externality, since the polluter does not need to bear the cost of it. The horizontal curve is the price curve, representing the amount that the producer can charge for his product, which is set by the market. According to supply-and-demand theory, the producer will fix his production where the price curve intersects with his cost curve, the point indicated as the 'original output' on the figure. If he produces more than this, he will pay more for the output than he will gain in revenue. If he produces less then he could be more efficient, based on economies of scale, by producing and selling more.

If a Pigouvian tax is introduced, the cost curve will shift upwards away from the x-axis, so that at each point the price per quantity produced is higher by an amount equivalent to the tax. This higher curve is the social

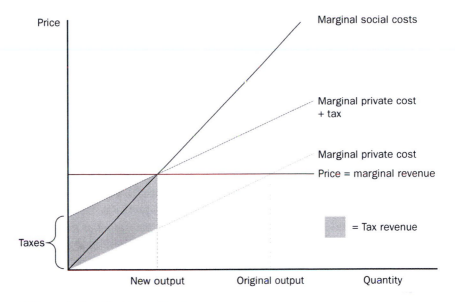

Figure 3.2 *Setting a Pigouvian fee*

Source: Drawn by Imogen Shaw

cost curve and labelled as marginal private cost + tax, since it represents the cost of production as well as the costs to society of by-products of the production process. Hence, the externality of pollution is internalized in the sense that it now features in the firm's accounts as a cost. As a result of the tax, the producer now faces the full cost of production and will produce a lower quantity, as indicated by the 'new output' point on the graph. The grey shaded area represents the revenue from the tax. According to economic theory, simply imposing this tax will rectify the market failure represented by the pollution; there is no need to pay the revenue to those who suffered from the effects of the pollution.

In this example, we have seen that society as a whole is recompensed, in the form of tax revenues, for pollution produced by an individual factory. In 1937, Ronald Coase approached the same problem in a different way, arguing that the problem of externalities can most efficiently be resolved as a market transaction between individuals: the producer who creates the pollution and the victim of that pollution. This solution, known as the 'Coase theorem', is based on the assignation of 'property rights' to the individuals, rather than finding a social or political solution involving government.

The Coase theorem suggests that, so long as a property right is assigned, the outcome of the negotiation between the two parties over the negative effects of the pollution will be efficient. It is usually demonstrated by means of an example. Hussen (2000) uses the example of two businesses – a paper mill and a fish farm – which share the use of a river. The paper mill is upstream of the fish farm, and releases a certain amount of effluent from its paper-making into the river, threatening the operation of the fish farm, which is situated downstream. The situation facing the two firms is presented in Figure 3.3. The MCC curve is the cost the paper mill will face in using other means of cleaning up effluent, rather than the river; the MDC curve is the marginal cost of damage caused to the fish farm by discharges from the paper mill. The natural equilibrium between the costs and benefits of the pollution, determined by the market, is at point S, where pollution is at level W_e. Coase argued that this will be the case regardless of who owns the property rights over the river, i.e. the power to decide to what extent it is polluted. In other words, it is unimportant whether the paper mill has the legal right to pollute the river, or whether the fish farm has the legal right to clean water.

If we first try assigning the ownership of the river to the fish farm, it would prevent all emissions from the paper mill (position 0 in the graph). However, if the mill were to discharge less than W_e of waste, the cost of alternative means of cleaning would be greater than the damage to the fish farm (MCC > MDC), giving the mill an incentive to pay the farm for the damage resulting from some level of pollution. There is a range of costs for this compensation (in the range from 0 to C_1 on the graph) representing the range of options where the marginal cost of alternative clean-up is greater than the damage to the fish farm.

The Coase theorem argues that, so long as property rights are clearly defined, it makes no difference who they are assigned to. It is the fact that

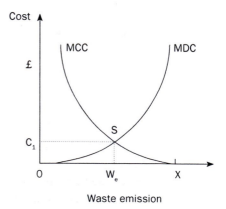

Figure 3.3 Illustration of the Coarse theorem

Source: Based on Figure 11.2 in Hussen, (2000) and redrawn by Imogen Shaw

somebody owns the river, and therefore has an incentive to protect it or charge for its pollution, that matters. This suggests that the same procedure of negotiation to achieve an economically efficient outcome would be possible if the ownership of the river had been assigned to the paper mill. In this case, the paper mill could discharge all its waste into the river, polluting the river to a level represented by the point X on the axis. But for all levels of waste between W_e and X (MDC > MCC), the paper mill would gain more financially by engaging in a negotiation to reduce its level of its emissions, and take a fee from the fish farm in return. So from this perspective also the optimum level of pollution is W_e, where MDC = MCC.

The Coase theorem is appealing to neoclassical economists because it reduces the role of government to merely assigning property rights, i.e. the rights to pollute the river or protect its purity; however, there are several flaws with it. As demonstrated in the example just given, it relies on a world where the origin of the pollution is clear – this is not the case with many of the most serious sources of environmental pollution. It can also be criticized for being totally pragmatic and not paying any heed to who is responsible for the pollution, thus flying in the face of the **polluter-pays principle**. The Coase theorem is based on the premise that it is irrelevant in efficiency terms where the right to pollute is assigned. However, this may matter very much indeed to the people who will suffer the health effects of the pollution, or who may lose their livelihoods because of that pollution. Perhaps most seriously of all, the Coase theorem was developed in a simple situation involving two producers and two possible victims, both living in the same jurisdiction. This can have little relevance to the most serious environmental problems that face us, which 'transcend national boundaries, involve irreversible changes and considerable uncertainty, and call for a coordinated, multifaceted response by a large number of nations' (Hussen, 2000: 231). Effectively, the negotiation and decisions costs of a private bargaining solution will rise rapidly the greater the number of stakeholders involved. In real life situations, this approach therefore becomes impractical.

3.4. Discounting the future

The consequences of many environmental losses and impacts are likely to be felt many years into the future. In the case of climate change, we may be talking about impacts that will be felt from 2050 to 2100; in the case of nuclear pollution, we are talking about radiation that will continue to be emitted into the environment for hundred of thousands of years. This represents a significant problem for economists whose techniques are based on markets and prices, since they need to be able to say what those prices are likely to be many years ahead. To achieve this, they use a technique known as 'discounting'. This translates the environmental impact from the future into a present value which is expressed as:

$$PV(B) = B_T/(1-r)^T$$

Here r is the discount rate, and B is the benefit or cost (C) accruing in T years' time. Such a formula has the effect of diminishing the calculated impact of environmental destruction caused in this present time period, and making our current actions appear less costly to future generations. The remainder of this section explains technically how this happens, but the important point to grasp is that these discount rates are applied routinely when the potential environmental impacts of developments are considered. So the values of damage that they imply – the accuracy of which depends on mathematical formulae and assumptions – determine whether such developments are damaging or not.

When working out the costs and benefits of any economic policy or production process over time, the outcome depends entirely on the discount rate that is applied. The higher the discount rate, the lower the future costs of current actions. Some economists favour an approach to discounting that is called 'descriptive', which assumes that the discount rate should just be equivalent to the prevailing interest rate. If interest rates are relatively high, e.g. 5–10 per cent, then the calculated present cost of any actions we take now that have negative environmental impacts will weigh very little in the future, since the discount rate has the effect of massively diminishing the present value of the distant future. The alternative approach to discounting is known as the 'prescriptive' approach, and is based on the sense that, although high rates of interest are potentially possible, people making long-term investments tend to choose less risky options, such as long-term government bonds, which have much lower returns, more in the range of 1–2 per cent or less above the rate of inflation.

Discount rates are calculated in two parts: the first part is labelled 'time preference'; the second is the 'wealth' component. The time preference element is based on the question of whether you would rather have something beneficial in the present or at some future date. This 'time preference' component is a source of much debate, since, at least within the course of a human life, it should be zero – jam today and jam tomorrow having equivalent **utility** value, in the economic jargon. However, experiments and everyday experience suggest that, in reality, people are impatient, and prefer to have things now rather than later, suggesting that they have what economists would call a 'positive time preference'. Somewhat ironically, the suggestion that we ourselves are undermining the possibility of future life for human beings on earth may actually greatly increase our time preference for present consumption. The wealth component is based on the assumption that incomes will rise, meaning that future generations will be richer than the present one. So if we are concerned with equity, we should do less to protect future generations, who we assume will be richer than we are:

> The source of the paradox is the assumption that future generations will be better off than we are; in this story, we are the poor, and those who come after us are the rich. If that were true, then as modern Robin Hoods we could strike a blow for equality by taking money from our children's inheritance and spending it on ourselves today.
>
> (Ackerman, 2009: 87)

Again, there is an obvious problem with this line of reasoning, since the idea of ever-increasing consumption is itself based on the economic growth that may be destroying the potential for future generations to enjoy their comfortable lives. In this sense, we might reasonably suggest a negative wealth component to the discount rate.

At the extreme, you might wish to argue that the only legitimate discount rate is zero, since all generations' preferences should be treated equally, and the period in which somebody lives should not affect their right to an equal quantity of well-being. If we take a more conventional economic view, that damaging impacts today will be of less importance in the future, we can begin to try to calculate what rate of discount might be appropriate. The rate we choose is dependent on a range of variables: our preference for consuming today rather than tomorrow; how much our consumption is worth to us (which relates to how much we already have, so it is higher for poorer people and countries); and the rate at which national consumption is increasing (which relates also to population growth rates).

Table 3.1 gives estimates made by the World Bank for national discount rates in 1990. You can see that the discount rates are negative for poorer countries, which means that as time goes by the future value of what happens now is highly valued in such societies. This suggests that, for these countries, future consumption is valued more than present consumption, which implies that these countries should have very protective attitudes towards the environment. This is not found in reality,

Table 3.1 Estimated discount rates for a range of countries, 1965–88

Country	Growth of real private consumption (1)	Growth of population (2)	Discount rate (%) (1 – 2)
USA	3.3	1.0	+2.3
UK	2.8	0.2	+2.6
Japan	5.0	1.0	+4.0
Ethiopia	2.4	2.8	–0.4
Ghana	1.7	2.6	–0.9
Chile	0.8	1.7	–0.9
Thailand	5.8	2.5	+3.3

Source: World Bank (1990)

since poorer countries are the sites of some of the worst environmental destruction, which suggests a flaw in the theory of the discount rate. The rates for developed countries are high, which, if they were applied to environmental problems, would mean that we would make little effort to protect the environment – since the discount rate would suggest that, not so far into the future, the impact of our present behaviour would have been greatly diminished.

While this may seem an arcane and technical discussion, it has an enormous impact on our chances of protecting our environment. These discount rates are applied when future impacts of current policies are calculated, and if the equations are in error then we risk huge future damage to our environment. How this has impacted on the problem of climate change in particular is discussed further in Chapter 13.

3.5. Case study: SO$_2$ allowance trading

'Acid rain' was one of the first environmental problems that could be scientifically proved to be caused by specific pollution processes, and was therefore an important test case for how such harmful pollutants might be controlled on a national scale. Acid rain is caused by the dissolution of pollutants from the burning of fossil fuels (especially in power stations) in rain or other precipitation. Although it was first identified in the nineteenth century, it increased in spread and intensity as the burning of fossil fuels expanded. The most significant gas that causes acid rain is SO$_2$, or sulphur dioxide. This case study explores the consequences of a market-based policy, framed within the neoclassical paradigm, to control SO$_2$ emissions introduced in the USA.

The scheme was introduced following the passage of the Clean Air Act amendments in 1990. Title IV of these amendments established an allowance trading programme for SO_2; its aim was to cut emissions by 50 per cent, or some 10 million tonnes, by 2000. The scheme worked by allocating to electricity-generating plants the right to produce sulphur dioxide, and allowing them to trade their quota of rights between themselves:

> Individual emissions limits were assigned to the 263 most intense SO_2 emissions-generating units at 110 electric utility plants operated by 61 electric utilities, and located largely at coal-fired power plants east of the Mississippi River. EPA [the Environmental Protection Agency of the US government] allocated each affected unit, on an annual basis, a specified number of allowances related to its share of heat input during the baseline period from 1985–87, plus bonus allowances available under a variety of provisions . . . Cost-effectiveness is promoted by permitting allowance holders to transfer their permits among one another, so that those who can reduce emissions at the lowest cost have an incentive to do so and to purchase permits from those from whom reducing the cost would be greater. Allowances can also be 'banked' for later use.
>
> (Stavins, 1998: 70–1)

From a neoclassical perspective, a trading scheme can achieve an efficient outcome because those who can most easily and cheaply reduce their emissions will do so, and will sell their permits to those whose production processes mean that it would be more costly for them to reduce emissions. The whole scheme works within a cap, which is determined by policy-makers based on the best available scientific advice. So an absolute limit is imposed, but the market decides efficiently who reduces emissions; thus the least-costly solution to the problem is found. Even within the neoclassical paradigm, there are problems with such a trading scheme. Rights are allocated on the basis of past polluting behaviour, so the heaviest polluters are given the greatest allocation (this is a process commonly known as 'grandfathering'). This creates what economists call a 'moral hazard', since polluters may be expected to resist cleaning up their act, or even polluting more, if they expect that a trading regime may be introduced. In the case of the SO_2 scheme, it appears that the moral hazard was minimized: allowances were granted on the basis of baseload contribution to the grid, rather than emissions, so that more efficient plants were not penalized. More radical economists would question the allocation of the value generated by the permits. If a right to permit any pollution is created and allocated to companies who can then sell it, then a right to pollute, which has been created by social convention and by decisions of politicians and policy-makers on behalf of the public, has been privatized and sold. Yet the value of this right, which was public, has been given to private companies free of charge – and they can profit from selling it.

The practical outcome of this scheme was impressive. The targets that were set for reductions were achieved and even exceeded. However, this in itself led to a problem, since companies were able to 'bank' a large number of allowances, raising questions about whether the limits originally agreed were sufficiently stringent. The US government's aim to achieve a reduction of 10 million tonnes in SO_2 emission by 1995 was achieved and exceeded. It was also achieved at less cost than a comparative regulatory solution: estimates have been made of the cost savings generated by the scheme, a figure of US$1 billion annual savings has been quoted.

Although such schemes may indeed create efficiency savings, it is hard to measure these in practice, because it is a difficult exercise to value something that never happened, i.e. the reductions that might have been brought about through a system of regulation. The damages caused by pollution – especially in the case of pollution that is deadly to human health, and whose results express themselves over many years and even generations, as is the case with radioactive pollution – are extremely difficult to measure accurately. The process of setting limits is also fraught with difficulty, and subject to political pressure from both environmentalists and those lobbying for companies that generate the pollution.

Market solutions are popular with neoclassical economists and polluters: the former because they have faith in markets; the latter because they gain the value of them. Taxes tend to be more popular with policy-makers and environmentalists: the latter tend to distrust the corporations who favour market solutions; the former receive the revenue from taxation. So long as the limit on pollution that is fixed initially is strict, and is based on scientific findings that are as immune as they ever could be from business lobbying, and so long as the value generated by emissions trading schemes becomes public rather than private property, even the more radical environmental and ecological economists see a role for market-based solutions in promoting efficient responses to pollution control. Other green and anti-capitalist economists challenge philosophically the 'enclosure' of the planet's atmosphere or water courses and their sale to the highest bidder (see more in Chapter 14).

Summary questions

- Why is scarcity important to a neoclassical economist?
- How likely is a CBA of building a new runway at Heathrow airport to take into account all the costs and benefits to all stakeholders?
- Can you think of other environmental problems where the example of sulphur dioxide emissions trading might be a useful guide for policy-makers?

Discussion questions

- What externalities can you identify in the business model of a leading supermarket?
- How indicative of its scarcity is the price of a particular non-renewable resource?
- What discount rate would you use when assessing whether or not to build a new nuclear power-station?

Further reading

Coase, R. H. (1960), 'The Problem of Social Cost', *Journal of Law and Economics*, 3/1: 1–44: a seminal work in neoclassical economics, addressing specifically the question of pollution.

Hanley, N., Shogren, J. F. and White, B. (2001), *Introduction to Environmental Economics* (Oxford: Oxford University Press), ch. 4 on cost–benefit analysis: an excellent, up-to-date introduction to the field of environmental economics, whose discussion on CBA is especially helpful.

Lawn, P. (2010), 'On the Ehrlich–Simon bet: Both were unskilled and Simon was Lucky,' *Ecological Economics*, 69: 2045–6: a brief summary of the famous wager and the limited understanding of its adversaries.

Simon, J. L. (1980), 'Resources, Population, Environment: An Oversupply of False Bad News', *Science*, 208: 1431–7: a short article which nicely conveys the attitude of the more extremely orthodox economics establishment to the ecological crisis.

⬤4 Environmental economics

The previous chapter focused on the way that the dominant economic paradigm has responded when confronted with environmental problems. To some extent, environmental economics works within the same paradigm and accepts many of the techniques and tools that neoclassical economics has developed. Environmental economists concern themselves with two main issues that arise from the recognition of planetary limits: environmental pollution and the depletion of scarce resources, including species. Environmental economics was initially a distinct sub-field of natural resource economics – which had a longer pedigree – but the two are now generally studied together.

As its name suggests, environmental economics foregrounds concern for the environment, and takes these issues most seriously: 'Environmental economics, in which environmental goods and services, as well as environmental risks, are given a monetary value, is the first systematic attempt to introduce the environmental dimension within mainstream economics' (Barry, 2007: 239). It may seem surprising that economists took so long to wake up to the fact that the economic system was running up against planetary limits; however, if we think back to the spaceman vs. cowboy metaphor that was introduced in Chapter 2, we can begin to understand how the thinking of economists prevented them from recognizing the limits – perhaps until it was too late.

The following section provides an introduction to environmental economics as the first branch of economics that has systematically prioritized the environment in its study. Section 4.2 explores a key aspect of environmental economics: creating markets for environmental goods, and especially how these might be priced. Section 4.3 explores a theory about how economic growth and environmental quality might be related. Section 4.4 considers how far markets are capable of protecting the environment, while Section 4.5 provides a case study of pricing a key environmental resource: the value of a British coastline.

4.1. Economics with the environment at its heart

Nick Hanley and his colleagues begin their useful *Introduction to Environmental Economics* (2001) by listing the insights from economic theory that they think are helpful to policy-makers seeking to protect the environment. These are reproduced in Box 4.1. This list indicates clearly that environmental economics is a school of economics, first, and then takes the environment into its thinking, although the authors also concede that there are insights from ecology which economists ought to be aware of.

Box 4.1

Ten key insights from economics which policy-makers need to be aware of

1. Economic and environmental systems are determined simultaneously.
2. People make decisions in response to incentives and to maximize utility.
3. Environmental resources are scarce.
4. Markets are the best way of allocating a vast range of resources.
5. Environmental problems arise from market failure.
6. Government intervention can make things worse.
7. Environmental protection costs money.
8. When managing renewable resources, choosing the maximum sustainable yield is rarely optimal.
9. Economic growth is not a panacea, but has vastly improved the quality of life of most people.
10. Environmental problems are global, and negotiating solutions through agreements will be hard.

Source: A summarized version of a list in Hanley et al. (2001: 9)

We can see from Box 4.1 that environmental economists have faith in the market as a useful allocation mechanism, and believe that the reason it has generated so many environmental problems is due to 'market failure', i.e. the market not operating as efficiently as it should. As we will see later, their proposed solutions revolve around various techniques to ensure that markets can take into account environmental costs and benefits. Environmental economists use the methods of conventional economics – with its reliance on mathematics – in both analysing the problem and seeking solutions. Hence, they will use a diagram such as Figure 4.1 to explain how economic activity and waste are related.

The figure makes clear that environmental economists take the natural limits of the planet very seriously: 'First, like anything else in nature, the assimilative capacity of the environment is *limited*. Thus, the natural environment cannot be viewed as a bottomless sink. With respect to its capacity to degrade waste, the natural environment is, indeed, a *scarce resource*' (Hussen, 2000: 92). Figure 4.1 assumes that there is a positive linear relationship between waste and economic activity, i.e. that as economic activity increases, the amount of pollution increases at a proportional rate. The 45° line in the figure is a visual representation of this relationship and the equation $W = f(X, t)$ indicates that pollution (W) is a function of the level of economic activity (X) and the variable t, which represents technological and ecological factors. If we assume that the latter are fixed, then we can conclude that the environment has a fixed assimilative capacity, which is represented in the graph as the dotted horizontal line which intersects the y-axis at W_0. Hence a level of pollution represented by X_0 can safely be absorbed.

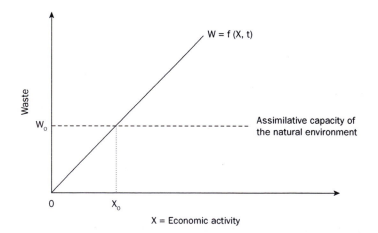

Figure 4.1 *Illustration of the relationship between waste emissions from production and the assimilative capacity of the environment*

Source: Based on Figure 5.1 in Hussen (2000) and redrawn by Imogen Shaw

If we relax our original assumption, we could model a change in the level of technological sophistication with which we deal with pollution by moving the dotted line upwards, meaning that the environment could now assimilate more pollution. Alternatively, we could change the relationship between the level of economic activity and the rate at which pollution is discharged, perhaps by moving towards more efficient production processes. This would be represented by the diagonal line moving downwards so that it was less steep, so that again we could increase the level of X without further degrading the environment.

The primary aim of environmental economists is to protect the environment, but they simultaneously hold strongly to the belief that markets are powerful mechanisms that can be used to safeguard the environment against the potential negative effects of economic activity. The environmental crisis makes it clear that markets are currently not protecting the environment, and so environmental economists seek to explain this 'market failure':

> Market failure comes about when people cannot define property rights clearly. Markets fail when we cannot transfer rights freely, we cannot exclude others from using the good, or when we cannot protect our rights to use the good. Under these conditions, free exchange does not lead to a socially desirable outcome because we either provide too much of bad goods like pollution or too few of good things like open space.
>
> (Hanley et al., 2001: 16)

Market failure is central to environmental economics: because environmental economists consider that the best way to protect the environment is to create a market, they are concerned, first, to establish a clear system of property rights, so that each person knows what it is in their individual interests to protect. Once this system is established and the market in environmental goods and services can be created, environmental economists must find ways to commodify and price the environment. The following section addresses how they create prices for non-market, environmental goods. (There is a discussion of market failure with reference to climate change in Section 13.2.)

Hanley et al. have three central explanations for market failure, which are interrelated:

1. *The* **externality** *problem*, i.e. the fact that pollution is 'external' to the operation of the factory or business, and hence can be ignored in the company balance sheet (this is discussed further in Section 11.1). The market model can work only when all aspects of production are included in the equation. If a firm can find ways to increase its internal

'efficiency' by passing some of the costs of its production onto the wider community, for example through releasing pollutants rather than cleaning them up, then the model fails.

2. *The public goods problem.* Goods from which we all benefit, but which some of us pay for, or which damage us all but only some of us produce – what economists call 'public goods' – are not naturally 'market goods'. Economic theory defines public goods as those which are non-rival and non-excludable, i.e. our use of them does not reduce the ability of others to use them, and once they are provided we cannot prevent others from using them. It may be that some of the environmental public goods, such as clean air or biodiversity, need to be protected by centralized expenditure derived from taxation. Yet those who will pay these taxes may not value the public goods sufficiently, or have a real sense of their value, and thus will resent having to pay for them, and hence a market economy may struggle to allocate them efficiently. So a market system may result in too much pollution and too few footpaths. (There is a longer discussion of the nature of 'public goods' in Section 14.1.)

3. *The common goods problem* is related to the first problem, although it is concerned with goods which anybody can have access to, sometimes also called 'open access resources' (Chapter 14 considers this issue in detail). When it is unclear who owns a part of the environment, nobody has the right incentive to protect it. Environmental economists refer to this as a 'missing market', and seek to create a commodity that can be owned and therefore protected. Perhaps the most important of these common goods is the global atmosphere itself, which can help to explain why negotiations to reach an agreement to protect it have been so intractable.

While we may feel that in some sense the earth's resources should be common property, it is frequently some of the world's poorer nations that have the best-preserved ecosystems and are supporting the global environment; one example is the rainforests, which act as massive sinks for carbon dioxide and huge reserves of species. So creating markets whereby the richer Western nations could somehow compensate these poorer nations for maintaining these habitats might also help to address global inequality. Since 1992, the UN Conference on Environment and Development (UNCED) has been exploring ways to create global markets for environmental goods in a process that recognizes the valuable 'good' that the country providing the environmental service is offering the world. If global production processes despoil these environments then the countries that lose them should be financially compensated. Table 4.1

Table 4.1 Potential global markets for 'environmental goods and services'

Type of mechanism	Compensating benefit to host country	Global environmental benefits
Global markets		
Intellectual property rights/ **bio-prospecting** deals	Contracts and up-front payments to share any commercial returns from pharmaceutical and other products	Biodiversity, protected areas
Joint implementation/ **carbon offsets**	Foreign capital investment in energy and land use sectors	Reducing greenhouse gases, **carbon sinks**, biodiversity
Debt-for-nature swaps	Purchase of secondary debt in exchange for protected areas	Biodiversity, carbon sink
Market regulation/ trade agreements	Premium in importing markets for sustainable exploitation of resources	Biodiversity, wildlife, forests
Transferable development rights	Landowners/developers are compensated with alternative rights to develop areas with less environmental value	Biodiversity, protected areas, carbon sink
International compensation		
Global environmental facility	Payment of the incremental cost of conserving any global benefits	Biodiversity, protected areas, **ecosystem services**, carbon sink, international waters, reducing GHG emissions
Global overlays	Modifying conventional cost–benefit appraisals of projects to account for any global benefits	Carbon sink, biodiversity
Environmental funds	Long-term financing of environmental and community-based conservation projects	Biodiversity, protected areas, regional and trans-boundary benefits

Source: Pearce and Barbier (2000)

lists some of the proposed global markets that might be created – these are sometimes referred to as 'shadow markets' because they do not really exist.

4.2. Valuing the environment

The central task of the environmental economist is to create the missing markets, the absence of which they see as the primary cause of environmental destruction. The market requires commodities, people prepared to pay for them, and a price at which they can be exchanged:

> A critical step in the economic calculus is that between a preference for something and a willingness to pay to secure it. That is how markets work, and in affording economic values to environmental assets, functions and processes the economist is taking what are often, but far from always, non-market phenomena and stimulating willingness to pay for those phenomena.
>
> (Pearce, 1998: 14)

Pearce developed the concept of total economic value (TEV), which he considers to be made up of three economic functions: supplier of resources, assimilator of wastes and direct source of utility in terms of enjoying the view or feeling spiritually uplifted (ibid.: 41). He also distinguishes between four different types of 'value' that are provided by the environment:

1. *Direct* values relate to resources that can be physically extracted from the ecosystem and then sold, or made into products that can be sold – examples might include wood from rainforests, plants that can be turned into medicines, and so on.
2. *Indirect* values relate to other 'services that the ecosystem provides but do not have a solid physical existence' – examples are the ability of certain plants to absorb chemical wastes and break them down, or the capacity of the earth's environment to absorb carbon dioxide.
3. *Option* values is the term that is used to describe money that people are prepared to pay to protect the environment so that they can derive either direct or indirect value from it in the future.
4. *Existence* values are an attempt to put into monetary terms the intrinsic value that people accord to the survival of an ecosystem in its own right, perhaps because they appreciate the view, or value the survival of species that rely on the ecosystem for their continued existence.

The first three types of value are all use values, whereas the fourth is a non-use value.[4]

Table 4.2 presents Pearce's summary of the TEV of a rainforest, based on a study of the the Korup National Park in southwestern Cameroon, which is funded by the Worldwide Fund for Nature. The rainforest is a rich and varied habitat:

> It contains Africa's oldest rainforest, over 60 million years old, with high species endemism. There are over 1,000 species of plants, and 1,300 animal species including 119 mammals and 15 primates. Out of the total listed species, 60 occur nowhere else and 170 are currently listed as endangered.
>
> (Pearce, 1993: 17)

Through its use as a tourist attraction, the National Park is able to generate income and receive government protection.

To create a value for these indirect environmental goods, economists create 'shadow prices', imputing prices to pseudo-goods that do not actually exist. Such prices are devised through a research process, for example by surveying people to determine how much they would be prepared to pay for the preservation of an environmental good, if it could be traded and if a market for it existed. There are fundamental problems with trying to price environmental protection: perhaps the most important problem is that the environments of countries where people have less money will automatically acquire a lower monetary value – not because

Table 4.2 Total economic value of a rainforest

Direct value (1)	Indirect value (2)	Option value (3)	Existence value (4)
Sustainable timber			
Non-timber products	Watershed protection	Futures uses under (1) and (2)	Forests as a source of spiritual value
Recreation	Nutrient cycling		Cultural and heritage value
Medicine	Climate change mitigation		Inheritance to future generations
Plant genetics	Micro-climate		Existence of species who rely on the ecosystem
Human habitat			

Source: Pearce (1993); Ruitenbeek (1990, 1992)

they are less valuable in any general sense but merely because the people living there, who would suffer if their environment were destroyed, cannot afford to offer so much to pay for it. Dresner cites the example of a disagreement between economists working for the Intergovernmental Panel on Climate Change (IPCC) and the Global Commons Institute, a London-based lobby group. The argument focused on the differential valuation of land in rich and poor countries. The loss of land in the countries of the South was valued at one tenth of the rate of the land in rich Western countries; the cost of a human life was similarly diminished:

> Based on an assessment of 'willingness to pay', the IPCC economists had valued the cost of a lost life in Western countries at US$1.5m. . . . They had valued a life at US$100,000 for the rest of the world. This is just one of a number of moral and practical criticisms that are made of the techniques of shadow pricing.
>
> (Dresner, 2002: 111)

There are a number of actual techniques that can be used to create pseudo-prices for aspects of the environment. According to Pearce, all have two stages: first, the economic value must be demonstrated and measured; second, it must be captured or 'appropriated'.

Market pricing techniques

The most straightforward way of assigning a market price to environmental damage is to see what the market costs associated with it actually are. Hence, if the environmental problem we are considering is the destruction of a crop by a pollution incident, then the farmer can simply be given an amount of money equivalent to what he would have received had he sold his crop. In a variation of the same method known as the 'replacement cost approach', the cost of a pollution incident can be measured as the cost of restoring the environment to its state before the incident occurred. For example, if a watercourse is polluted by a factory, the price to be charged to the polluter is whatever it would cost to clean up that pollution. So, in a cost–benefit calculation, this value can be included as a cost.

Household production functions

This technique is so named because it begins the process of arriving at a market value for an environmental good by considering households' expenditure on goods and services. The method involves costing the substitute that can be offered to the consumer who has lost out because something he or she values in the environment has been destroyed; we examine what individuals buy to prevent loss or substitute for what has been lost, and we use the cost of this to stand for the cost of the loss. One example might be the amount a person would spend on installing insulation to prevent noise from aircraft destroying the peaceful enjoyment of the view from their conservatory. The insulation was actually bought in a shop, and we can use its cost as a measure of the amount that the person valued the peace and quiet that they had before the runway was built near their home. A simpler example might be to value the loss of a park because of the development of a supermarket as equivalent to the cost of travelling to a park that is far from a person's home.

Hedonic price methods

The value that people place on the environment can be assessed implicitly from their market choices: because a house with a sea view will be worth considerably more than a house a few streets further away from the coast, we can ascertain that there is a value associated with living right next to the sea. Hedonic pricing is based on this principle, and involves assessing how much certain developments cost by measuring market changes. For example, in the hedonic housing market, the environmental impact of an airport runway being constructed can be assessed by the fall in local house prices that occurred once planning permission was given – or even sought. The price that exists in the real market is considered to be an implicit price for the missing market. Hedonic pricing can also be used to put a price on human life, by considering actual data on how much people spend on medical procedures compared to the wages they earn. Hussen (2000: 296–7) gives the example of a US factory on Lake Superior whose asbestos-related pollution was estimated to cause an additional 274 deaths over 25 years of operation, thus reducing life expectancy by 12.8 years for those living around the lake. The social cost of each death – based on their loss of economic productivity because of their early death – was valued at US$38,849, which allowed those considering whether to allow the continued operation of the plant to estimate the social cost from its pollution at more than US$10 million.

Experimental methods

The previous methods are all conducted by environmental economists working from existing data and in the quietude of their offices. In experimental methods, they venture into the world and discover how much people value aspects of the environment by asking them directly what they would be prepared to pay to protect it. In a method known as 'contingent valuation', people are asked what they would be willing to pay to protect their local park or to avoid having a nuclear power station built in their community, for example. The method known as 'contingent ranking' or 'stated preference' considers how much people value an environmental good relative to other goods that are actually bought and sold in a market, thus enabling the researcher to fix the relative price of the environmental good that they are interested in.

It is clear that these various techniques are hugely complicated (and expensive) to use, and that the prices they arrive at can never be considered to have a definite relationship with the value that people place on the environmental good or resource that is under threat or has been lost. An environmental economist would argue that, in a society where markets dominate, pricing the environment – no matter how inadequately – affords the environment the best protection. Critics might suggest that a more pragmatic conclusion would be that some areas of life are too precious to be included in the sphere of the market.

4.3. When will we be rich enough to save the planet?

The previous section indicated that, if we create pseudo-markets for environmental goods, then those in poorer countries – or poorer areas in wealthier countries – will find their environments less well protected. Another way of looking at this issue is to consider development as a movement towards the increasing demand for more sophisticated post-material goods. Poorer people are struggling to meet their basic needs for food and shelter, and so cannot afford to concern themselves with protecting their environment. Environmental economists suggest that there is an inverted U-shaped curve relating income to indicators of environmental quality. They call this the 'environmental Kuznets curve' (EKC, see Figure 4.2), referring to a similar-shaped curve that Simon Kuznets used to described the relationship between income levels and equality within a society. The curve implies that, while development may initially result in poorer environmental standards as pollution levels rise, when countries become richer still they begin to prioritize environmental quality, and hence it rises again. For example, 'We find that while increases in GDP may be associated with worsening environmental conditions in very poor countries, air and water quality appears to benefit

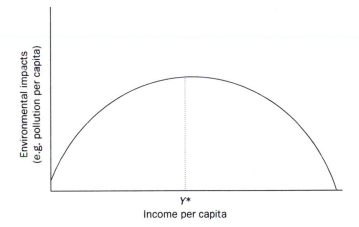

Figure 4.2 *Environmental Kuznets curve*

Source: Drawn by Imogen Shaw

from economic growth once some critical level of income has been reached' (Grossman and Krueger, 1991: 18–19); this level is indicated by $Y*$ in the figure. More succinctly, the evidence suggests that it is possible to grow your way out of environmental problems.

As Ekins notes, these are strong conclusions:

> They create the impression that economic growth and the environment are not only not in conflict – the former is necessary to improve the latter. They invite an emphasis on achieving economic growth rather than on environmental policy, because the former is perceived to be able to achieve both economic and environmental objectives, while the latter may impede the former.
>
> (Ekins, 2000: 183)

The state of the environments of the world's richer nations can itself be used as evidence against the theory of the EKC: if we are not rich enough by now to protect our environments, when will the turning point arrive? And can we find enough resources and energy on a limited planet to reach that point?

Ekins concludes that the evidence for the EKC is mixed, and varies depending on the type of pollutant we are considering. Evidence for the EKC hypothesis for environmental quality as a whole is not convincing – the evidence is strongest in the case of air pollution, including carbon monoxide, nitrous oxide and sulphur dioxide. However, there is no way of determining whether this relationship is causal, and it does not appear to arise automatically from economic processes but rather from political decision-making: 'Any improvements in environmental quality as incomes increase is likely to be due to the enactment of environmental policy rather than **endogenous** changes in economic structure or technology' (Ekins, 2000: 210).

The EKC hypothesis also fails to draw a distinction between different groups within societies. It may be that there are elites within poor countries who benefit from increased economic activity but who can remove themselves from the negative environmental consequences of that activity – for example, they may manufacture clothes in a factory that pollutes the local river, which impacts on the life of villagers downstream, but themselves live in a distant urban centre. So the villagers may demand pollution control, but will have no power to ensure it. It is fairly clear that the environmental regulation that has been introduced in the richer Western countries has been the result of determined and persistent political activity, so there is no automatic connection– and, for countries without democratic systems, perhaps no connection at all – between economic development and improved environmental protection.

4.4 Can markets save the planet?

So the central aim of environmental economists is to extend the market mechanism – and especially the price mechanism – to encompass the environment. Their efforts to create prices for environmental goods, and even to use this vocabulary, has been criticized by ecological economists, who question what exactly is being 'valued' and whether the monetary value that is arrived at truly reflects the full value of what may be lost. As we will see in the following chapter, many economists are concerned about this attempt to commodify certain fundamental aspects of the planet as **ecosystem services**. In responding to his critics, Pearce is keen to stress that it is actually an indication of his commitment to environmental protection:

> They are economic functions because they all have a positive economic value: if we bought and sold these functions in the market-place they would all have positive prices. The dangers arise from the mistreatment of natural environments because we do not recognize the positive prices for these economic functions. This is not the fault of economics or economists . . . Indeed, environmental economists have been at considerable pains to point out these economic functions and to demonstrate their positive price.
>
> (Pearce, 1998: 41)

Pearce is also keen to draw a distinction between measuring the value that people place on their environment and actually putting a price on the environment itself. He claims that environmental economists are attempting to do the former, and that they have been sorely misunderstood by critics who claim they are trying to do the latter. The environment has both an intrinsic value and an economic value, he claims, and using money as a measuring rod to assess the one does not diminish the other; in fact, it might provide a means of better protecting it (ibid.: 13).

This chapter has introduced the response of orthodox economics to the environmental crisis – changing the objective of economics without changing the methods used. For some economists, this is not going far enough: 'Environmental economics may be criticized for economizing the environment rather than ecologising economics' (Barry, 2007: 240). As we will see in the following chapter, the pricing of the environment is a key point of disagreement between environmental economists and ecological economists, who claim that some aspects of life are literally priceless, and that attempts to create pseudo-prices for them may miss the point.

4.5. Case study: How much do you like to be beside the seaside?

Approaches to valuing and protecting coastline resources have undergone considerable change over the past two decades, as extreme weather events resulting from climate change have exacerbated traditional coastal erosion. In response, local authorities in the UK have been required to draw up Shoreline Management Plans (SMPs), which include an assessment of the way that local people and visitors value their shorelines. The purpose of these strategic documents is to manage coastlines before problems with flooding or erosion become acute. They are required to consider four possible options: do nothing; hold the existing coastal defence line; advance the existing line of defence; or bring the existing line of defence further inland.

Poole Borough Council, on the south coast of England, is responsible for a stretch of coastline several kilometres in length, which is widely used for bathing and recreation by locals and tourists, and includes the prominent seaside resort of Bournemouth as well as Poole itself. The coastal mudflats, salt marshes and reed beds also offer habitats for a variety of aquatic birds. As a contribution to developing its SMP, Poole Borough Council conducted a valuation exercise based on the contingent valuation method. Researchers sampled 470 people who were on the beach over a number of days in the summer of 1995; these included locals, day visitors and resident visitors. They questioned them about a variety of issues, including how much they would be prepared to pay to protect various aspects of the experience of spending time on the Poole coastline, how likely they would be to spend their time elsewhere if the coastline were eroded, and how much this would cost them (the results are reported in Table 4.3).

The results in the table indicate that there would be significant losses in the value of people's enjoyment as a result of erosion. While all the beach users could find an alternative beach, they would not value it as highly, and would also face considerable additional costs in reaching it. The authors concluded that if the coastline were not protected then there would be a loss of some £131 million in recreational benefits over 50 years.

Table 4.3 The value of enjoyment related to the level of erosion of the coastline at Poole (mean values, £)

Type of user	Before erosion	After erosion	Alternative site	Cost of transfer
Locals	7.60	3.20	6.30	2.20
Day visitors	11.30	3.90	9.50	1.80
Residential visitors	13.10	5.70	11.40	2.40
Total	10.60	4.30	9.10	2.10

Source: Polyzos and Minetos (2007)

Summary questions

- What do environmental economists mean by market failure?
- The EKC would suggest that poorer countries always protect their environments less well than richer countries. Do you think this is actually the case?
- Think of an environmental loss that occurred in your experience. Which of the pricing methods described would best put a price on the loss you experienced, if any?

Discussion questions

- How can we price the value of rainforests in mitigating the worst effects of climate change?

- How useful would it be to include 'shadow prices' in international climate change negotiations?
- Given the complexity of the process of pricing the rainforest given in Table 4.5, do you think this is a productive investment?

Further reading

Hanley, N., Shogren, J. F. and White, B. (2001), *Introduction to Environmental Economics* (Oxford: Oxford University Press), especially ch. 3: an excellent and timely introduction to the field.

Hussen, A. M. (2000), *Principles of Environmental Economics: Economics, Ecology and Public Policy* (London: Routledge): a more technical introduction, including graphical and mathematical content.

Pearce, D. W. and Barbier, E. B. (2000), *Blueprint for a Sustainable Economy*, 2006 edn (London: Earthscan): a primer for the more generalist reader from one of the key pioneers of the field, together with a co-author.

Williams, J. B. and McNeill, J. M. (2005), 'The Current Crisis in Neoclassical Economics and the Case for Economic Analysis based on sustainable development', U21 Global Warning Paper: an account of the conflict between the traditional approach to economic development and environmental sustainability.

5 Ecological economics

While environmental economists have taken an important step forward in including the value of environmental resources in their economic model, they have not addressed the possibility that the very nature of their discipline, which begins with theory and mathematics and then applies these to the natural world, might be inherently incapable of protecting the planet. The ecological economists have taken a different tack, beginning with the science of ecology and seeking ways to link it to the discipline of economics. As well as having different origins, these two approaches to tackling the environmental consequences of economic activity differ in their definition of sustainability:

- Neoclassical environmental economists favour a goal of weak sustainability (technology will lead to **manufactured capital** substituting for **natural capital**) and have sought to adopt an objective stance.
- Ecological economists favour a goal of strong sustainability (meaning that physical capital cannot substitute for natural capital)[6] and are less concerned about preventing their personal viewpoint from impinging on their analysis.

This chapter explores the contribution of the ecological economists and the implications for our economic and political life. For the sake of simplicity, I have presented what appears to be a unified account, although it is important to note that there has been evidence of a 'conflicted movement' – with differences of opinion especially concerning the relationship with environmental economists, and a stronger emphasis on political economy amongst European as opposed to US ecological economists (Spash, 2009). We begin in Section 5.1 with an introduction to ecological economics as distinct from the neoclassical economics that it has reacted against. Then, in Section 5.2, I outline some of the key concepts and theories developed by the ecological economists, particularly those that have brought wisdom from ecology into economic discussions. Section 5.3 examines one of the key insights of this school: the importance of the steady-state economy (SSE). We conclude in Section 5.4 with a case study that helps to illustrate the underlying philosophy of ecological economics: a discussion of the viability of pricing aspects of our natural environment that environmentalists think are, in reality, priceless.

5.1. A break with tradition

Ecological economics developed in response to generalized concern with the state of the environment from the late 1960s onwards. A leading figure who was instrumental in developing this approach to economics was Herman Daly, whose work will be mentioned throughout this chapter. He was in turn influenced by Nicolae Georgescu-Roegen and Kenneth Boulding (whose contribution is discussed in Chapter 2). The academic subdiscipline of ecological economics was born in the late 1980s, with the International Society for Ecological Economics founded in 1987, and the launch of the journal *Ecological Economics* in 1989.

In their *Introduction to Ecological Economics*, Costanza et al. identify and rank the problems that they seek to resolve:

> First, establish the ecological limits of sustainable scale and establish policies that assure that the throughput of the economy stays within these limits. Second, establish a fair and just distribution of resources using systems of property rights and transfers . . . Third, once the scale and distribution problems are solved, market-based mechanisms can be used to allocate resources efficiently.
>
> (Costanza et al., 1997: 83)

This key text within the ecological economics field is concerned primarily with the size of the economy and the fact that it should not expand beyond the limits that the planet can sustain. The second key criterion is the commitment to social justice. Only once these criteria are met are its authors prepared to concur with the neoclassical economists that the market can be allowed the task of allocating resources. It is clear from this quotation that the language and style of ecological economics are academic in tone; it is committed to markets, efficiency and property rights, but makes strict conditions for the framework within which capitalist markets should function. In essence, it is seeking to create a form of market capitalism that does not threaten the health and well-being of the planet, but it makes no radical claims for a fundamental restructuring of our economic philosophy.

Ecological economists are scathing about the neoclassical approach – even that taken by the environmental economists. They condemn the blithe optimism of the neoclassicals' reliance on technology to solve the problems of scarcity and waste:

> Most neoclassical economists assume that technological advance will outpace resource scarcity over the long run and that ecological services can also be replaced by new technologies. Ecological economists, on the other hand, assume that resource and ecological limits are critically important and are much less confident that technological advances will arise in response to higher prices generated by scarcities. This difference in worldview, however, does not prevent neoclassical and ecological economists from sharing the same pattern of reasoning.
>
> (Costanza et al., 1997: 69)

There are a number of 'watershed' topics that illustrate the divide between environmental and ecological economists. As discussed by Illge and Schwarze (2006), these are:

- The concept of human behaviour that each embodies: whether the economic actor is a 'rational economic man' (environmental) or a person who lives in balance with the environment (ecological);
- The way in which nature itself is valued, whether in monetary (environmental) or biophysical (ecological) terms;
- Judgements about the relationship between sustainable development and growth;
- The extent to which economics should be considered as a scientific study;
- Differing emphasis on issues of distribution and justice.

Box 5.1 outlines the vision of ecological economics as expressed by some of its leading proponents. Figure 5.1 illustrates this vision and provides a visual representation of the central demand of ecological economists: that the economy should be recognized as necessarily existing within the ecosystem.

Box 5.1

The vision of ecological economics

1. The vision of the earth as a thermodynamically closed and non-materially growing system, with the human economy as a subsystem of the global ecosystem. This implies that there are limits to biophysical throughput of

resources from the ecosystem, through the economic subsystem, and back to the ecosystem as wastes.

2. The future vision of a sustainable planet with a high quality of life for all its citizens (both humans and other species) within the material constraints imposed by 1.

3. The recognition that, in the analysis of complex systems like the earth, at all space and time scales, fundamental uncertainty is large and irreducible, and certain processes are irreversible, requiring a fundamentally precautionary stance.

4. That institutions and management should be proactive rather than reactive, and should result in simple, adaptive and implementable policies that are based on a sophisticated understanding of the underlying system and which fully acknowledge the underlying uncertainties. This forms the basis for policy implementation that is itself sustainable.

Source: Costanza et al. (1997: 79)

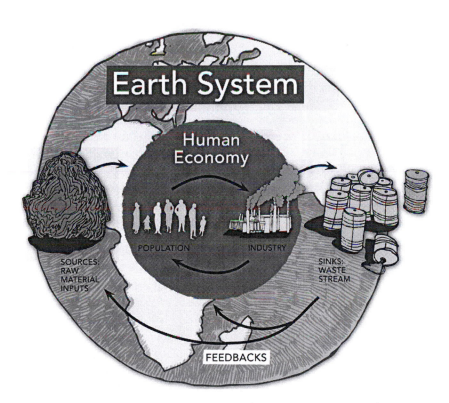

Figure 5.1 *The place of the economy according to ecological economics*

Source: Drawn by Imogen Shaw

5.2. Thinking differently about the economy

The contribution of ecology

Kirkpatrick Sale (2000) identifies four 'conditions of an imperiled environment' drawn from ecology, which are relevant to the current stage of human evolution:

- *Drawdown*: the process by which the dominant species in an ecosystem uses up the surrounding resources faster than they can be replaced.
- *Overshoot*: when the use of resources in an ecosystem exceeds its carrying capacity and there is no way to recover or replace what is lost.
- *Crash*: a precipitate decline in species numbers.
- *Die-off*: the extinction of species that cannot reorganize their ecological functioning following a crash.

Ecological economists take a scientific approach to studying our species and its inhabitation of an ecological niche, just as an ecologist would with any living species. The consequences of this approach when considering how we are treating our planetary support system are stark, and are particularly so when you realize that we are the only species that has ever been capable of chronicling its own ecological demise in this way. Ecological economists are also concerned to maintain a respect for the complexity and **holism** of natural systems, even when this makes the study of the economy fraught with difficulty, and prohibits the creation of economic models. This is the source of their divergence from the orthodox method of economic analysis (see Martinez-Alier, 2002).

Ecological economists use a range of key concepts drawn from the science of ecology to describe the interaction between nature and the economy. An example is the concept of assimilative capacity. This refers to the way that the environment can treat and absorb harmlessly a certain level of waste, even when this waste is actually polluting. For example, the atmosphere can absorb a certain amount of CO_2 without it increasing global temperatures; and plants can absorb and digest toxins, turning them into harmless chemicals. The ability of nature to reabsorb the outputs of production processes is sometimes – and controversially – referred to as **ecosystem services**. However, such 'services' are limited by the biological processes that underpin them. If the volume of pollutants, or the speed at which they are produced relative to the speed of natural processes, is too great, then the process of assimilation will be overwhelmed, and the very system that enabled the breakdown and reabsorption to take place may be destroyed. This is a very dangerous example of sawing off the branch of the tree we are sitting on:

> If we dispose of wastes in such a way that we damage the *capability* of the natural environment to absorb waste, then the economic function of the environment as waste sink will be impaired. Essentially we will have converted what could have been a renewable resource into an exhaustible one.
>
> (Pearce, 1998: 39)

By analogy, at the level of input rather than waste we have the concept of regenerative capacity. This refers to the ecosystem's ability to replace resources that we use in our production systems or to recover from the pollution they produce. We can divide resources between those that are **renewable** and those that are **non-renewable**. Renewable resources, such as wood or wind energy, are in continuous supply, although the rate at which they can be replenished will vary from resource to resource. Non-renewable resources, such as iron ore or fossil fuels, are in limited supply within the earth's crust, and thus they cannot be replaced once they are used up. A sustainable economy would obviously deal differently with renewable and non-renewable resources, and would take account of regenerative capacity with regard to any particular renewable resource.

If we take an ecological perspective on human activity then it is clear that we need to think about the niche we are occupying, how it meets our needs, and the number of us whose needs it can continue to meet. Thus it becomes clear that, from the perspective of carrying capacity, the issues of population size and resource use interact in a complex way. As a simple example, the planet could withstand the lifestyle of many more Chinese peasants than one US citizen, given the relative alacrity with which each uses resources and produces wastes. Holden and Ehrlich summarized this understanding in the equation $I = PAT$, where I is environmental impact, P is the absolute size of the population, A is affluence, or the average per capita consumption of the people making up the population, and T is the environmental impact of the productive technology of their society (Holden and Ehrlich, 1974; see the more detailed discussion of this in Section 2.3). This formula suggests that, although limiting population is an important contributor to living within the planet's means, rising consumption levels are a more serious threat.

Thus the IPAT equation draws attention to the importance of the intensity of human living as well as its scale. Schumacher's famous adage, 'small is beautiful', is usually interpreted as meaning that a sustainable economy would be designed on a small scale, but, according to Martinez-Alier and colleagues (2001), scale is shorthand for the combined effects of population and average resource use, a concept which they assign to Herman Daly. The conclusion is that, when considering resource use, we should think of scale in terms of human impact rather than the absolute magnitude of population.

Systems thinking and thermodynamics[7]

Many environmental scientists encountering the worldview of the economist for the first time are astonished by the blasé way in which economists deal with physical reality, and in particular with the laws of thermodynamics. Thermodynamics is among the most important topics in science: it studies how energy is exchanged between physical systems such as heat and work, resulting in changes in pressure, volume, temperature and entropy, the measure of disorder within a system. The laws of thermodynamics provide some of our most basic understandings of what is physically possible. Yet it was not until 1971 that Nicolae Georgescu-Roegen, one of the first economists to attempt to apply the laws of thermodynamics to the production of goods within an economy, published his classic *The Entropy Law and the Economic Process*. This initial disjuncture between economic theorizing and science is perhaps not so surprising once you realize that the foundations of economics pre-date an understanding of thermodynamics: the laws of thermodynamics were only formulated in the first half of the nineteenth century, whereas the basic postulates of economic theory were drawn up some 50 years earlier, with Adam Smith's *Wealth of Nations* published in 1776 and Ricardo's *Principles of Political Economy* published in 1817. Unfortunately, from an environmental point of view, much of subsequent economic theory has consistently failed to keep up with, or even acknowledge, our constantly improving scientific understanding of the physical universe within which economic activity takes place.

In fact, one of the major contributions of ecological economics is its emphasis on the importance of applying scientific understanding to economic theory. The first and second laws of thermodynamics in particular are seen as crucially important to understanding what any economic process can hope to achieve. The first law of thermodynamics, also known as the conservation law, states that energy can be neither created nor destroyed, only transformed. However, the second law of thermodynamics – 'the entropy law' – tells us that in general, the total amount of useful, organized energy available for work is always declining. For example, a lump of coal is a high-quality, highly organized form of energy; when burned, it turns into smoke and heat, which are low-quality, dispersed and much more disordered forms of energy. This process is irreversible: we cannot recapture all the heat and smoke produced by burning and turn it back into a lump of coal. The second law tells us that all energy systems have a tendency to increase their entropy (or degree of disorder) rather than decrease it. This appears to apply to everything in the physical universe. So, although many natural and technological processes increase order on a local scale – through the growth of plants, say, or the manufacture of goods from raw materials – the material waste and heat produced by these processes steadily, if imperceptibly, increases the general disorder of the physical universe. Even the stars themselves are part of this process – they will eventually burn all their fuel and disappear, leaving only homogenous regions of space where pressure, temperature and density are evenly and randomly distributed.

From the perspective of ecological economics, one of the most important connections between the first and second laws is the understanding that in any process in which energy is used – whether by machinery or living things – much of the energy involved will not end up in the goods or services produced. This means that we need to take seriously the waste products of our industrial systems. Kenneth Boulding argued that the economy currently operates on a linear system: we take materials and energy in at one end and simply throw away so-called 'waste products' at the other end. The valuable energy and materials contained in these resources are often simply thrown away into the environment, contributing toward an increasing disorder. An economy that took the laws of thermodynamics seriously would be a circular economy, where each waste product became the input into a new production process – an insight captured in Boulding's vision of 'spaceship earth'.

However, it is also important to remember that, in Boulding's phrase, 'we cannot turn pots back into clay'. In other words, once we have combined raw materials (e.g. clay) with highly ordered energy sources (such as the wood, coal or electricity used to fire the kiln) to create a sophisticated product, we cannot simply reverse the process to recover natural resources that can be used as an input to a new production process – any such recycling process will require (at the very least) the use of more highly ordered energy, can only be partially efficient, and thus will in turn create more waste. Currently, nature's ability to effectively reuse waste is vastly superior to ours. Ecosystems that have evolved over millions of years are typified by their ability to recycle energy and materials through complex food webs; new eco-conscious design movements such as permaculture and industrial ecology seek ways of designing production processes that mirror this ability.

Georgescu-Roegen's entropy hourglass analogy can be helpful in explaining the application of thermodynamic laws to economic systems (see Figure 5.2). Here is how Herman Daly describes the hourglass images:

> First, the hourglass is an isolated system: no sand enters, and no sand exits. Second, within the glass there is neither creation nor destruction of sand: the amount of sand in the glass is constant. This is of course the analogue of the first law of thermodynamics – conservation of matter-energy. Third, there is a continuous running-down of sand in the top chamber, and an accumulation of sand in the bottom chamber. Sand in the bottom chamber, since it has used up its potential to fall and thereby do work, is high-entropy or unavailable matter/energy. Sand in the top chamber still has potential to fall, thus it is low-entropy or available matter/energy. This illustrates the second law of

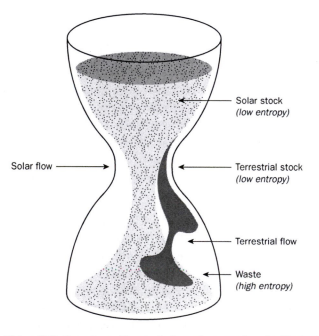

Figure 5.2 *Georgescu-Roegen's hourglass analogy to illustrate the entropy law*

Source: Drawn by Imogen Shaw

thermodynamics: entropy increases in an isolated system. The hourglass analogy is particularly apt because entropy is time's arrow in the physical world.

Source: http://www.eoearth.org/article/hourglass-analogy.

Mary Mellor's description of the capitalist economy as being like Oscar Wilde's *Picture of Dorian Gray* is another helpful illustration. In the story, Dorian Gray is a beautiful young man who engages in a variety of unspeakable and dissolute acts, but nonetheless remains physically uncorrupted. The twist in the tale is that he has a picture in his attic, which manifests the physical consequences of his dissolute lifestyle. Mellor argues that the apparent success of our economy in generating luxurious lifestyles and clever gizmos is actually bought at a price extracted out of sight and out of mind – a price that is paid by the world's poor countries and the planet itself. Taking the thermodynamic laws into account, ecological economists arrive at the conclusion that 'the economy really is an open subsystem of a materially closed, non-growing, and finite ecosystem with a limited throughput of solar energy' (Costanza et al., 1997: 173). They conclude that this basic reality of our physical environment should be respected in all economic planning and policy-making.

The second law has important implications for the policies that we need to adopt in response to the environmental crisis. The first move was to think about materials, and so recycling became a favoured policy option. But to recycle complex products, such as cars, which contain many different components and types of material, may ultimately use more energy than it saves:

> The serious problem of waste heat remains. The second law of thermodynamics tells us that it is impossible to recycle energy and that eventually all energy will be converted into waste heat. Also, it is impossible to recycle materials with one hundred percent completeness. Some material is irrecoverably lost in each cycle. Eventually, all life will cease as entropy or chaos approaches its maximum. But the second law of thermodynamics implies that, even before this very long-run universal thermodynamic heat-death occurs, we will be plagued by thermal pollution, for whenever we use energy, we must produce unusable waste heat. When a localized energy process causes a part of the environment to heat up, thermal pollution can have serious effects on ecosystems, since life processes and climatic phenomena are regulated by temperature.
>
> (Daly, 1971: 30–1)

Many bright ideas to protect the environment have foundered on the second law, which belies the possibility of any form of perpetual motion machine. The law similarly introduces realism into the enthusiasm for technofixes to the problem of climate change, such as high-tech solutions for capturing and containing carbon dioxide: it is unlikely that these can be developed without requiring so much energy as to make them pointless. In terms of industrial processes, taking thermodynamics seriously suggests that we should focus on designing them in ways that reduce economic friction (i.e. the amount of entropy that our productive processes generate) and enable the highest possible rates of reuse and recycling with the lowest possible energy input.

Extending the concept of capital

When Marx coined the term 'capitalism', he was thinking about one specific type of **capital**: the money that enabled those who owned and controlled it to dominate the allocation of economic and thus political resources. Capital as equated with money is a tool for facilitating economic activity; it does not represent any real value in itself. This is obvious, if you think about the real value of the pieces of paper we take to be money. But even at a deeper level, the value that money has is given to it by social and political agreement. However, within the discussion of various economists it seems to have taken on a life of its own, as though it has real value like that of land or labour. In fact, money's value is indirect: it represents the power to acquire goods, but many economists forget the fact that this power relies on social agreement. The acceptance or rejection of this redefinition establishes a clear dividing line between neoclassical, environmental and ecological economists, and the green and anti-capitalist economists, whose contributions will be considered in the following two chapters. The first three systems will establish capital as a value and then extend the concept to include other forms of capital than money, such as **social capital** or **natural capital**. Green and anti-capitalist economists are more sceptical about the role of money, and are concerned to keep it in its place.

Once capital is allowed to exist as a real entity in the economy, rather than as what Marx called an 'epiphenomenon' – something that is superficial to the real machinery of the economy – it becomes possible to argue both that we can substitute one form of capital for another (see Goodwin, 2007), and that we can substitute consumption in one time period for consumption in another (the problem of discounting, which is discussed in Chapter 3). This is a world of abstract theorizing that is detached from the reality of economic production, which uses the outputs of natural processes adapted through a combination of human and animal labour and energy. The UK-based think tank Forum for the Future has developed a model based on five different forms of capital, which has been described as follows:

> Behind the notion of capitalism lies the notion of capital – which economists use to describe a stock of anything (physical or virtual) from which anyone can extract a revenue or yield . . . When people think of capital in this sense, they usually think of some of the more familiar 'stocks' of capital: land, machines and money. But in the description of the Five Capitals Framework [. . .], this basic concept of capital (as in any stock capable of generating a flow) has been elaborated upon to arrive at a hypothetical model of sustainable capitalism. It entails five separate capital 'stocks': natural, human, social, manufactured and financial . . . The Five Capitals Framework requires a more **holistic** understanding of all the different stocks of capital on which our wealth depends [see Figure 5.3] . . . At its simplest, our wealth depends on maintaining an adequate stock of each of these types of capital. If we consume more than we invest, then our opportunities to generate wealth in the future will inevitably be reduced. **Sustainability** can only be achieved if these stocks of capital are kept intact or increased over time.
>
> (Porritt, 2009: 30–1)[8]

The possibility of substituting one type of resource for another is a source of conflict between ecological and environmental economists. An environmental economist would argue that it is efficient to compensate for the diminishing stock of **non-renewable** resources by increasing use of **renewable** resources. This sort of argument can sometimes drift into

Figure 5.3 The Five Capitals Framework

Source: Author's graphic

arguing that substitution between different types of capital might be possible, so that **natural capital** can be expended so long as this is compensated for by an increased level of physical or even **financial capital**. An ecological economist would call a halt at this point, feeling the need to defend the primacy of natural capital. While technical improvements can lead to more efficient use of resources, the substitution of natural capital by other types of capital is problematic. The conclusion from the ecological economists is thus that sustainable development requires careful stewardship of all forms of capital, and that this must take an equal place in economic calculations alongside considerations of maximizing productivity and **utility**.

5.3. From equibilibrium to steady state

The questioning of economic growth is so central to the economy–environment discussion that it has been given its own chapter in this book (Chapter 9). However, the discussion also merits a place in this chapter, since it is the central practical insight that results from the move taken by ecological economics to reposition the economy within the ecosystem: 'In the neoclassical view the economy contains the ecosystem; in the steady-state view the ecosystem contains the economy. This difference in view is rather like the difference between Ptolemy and Copernicus – is the earth or the sun the centre of the universe?' (Daly and Townsend, 1993: 3). As another commentator has noted, this is the central ideological shift taken by the ecological economists. It defines their discipline and all other insights arise from it: 'Ecological economics emphasizes that the planet is a thermodynamically closed, finite system, and hence environmental limits must be fixed prior to the occurrence of market exchange' (Greenwood, 2007: 78).

Theorizing about the importance of the SSE began with Herman Daly and his book *Steady-state Economics*, published in 1977. He built on the recognition of the importance of the second law of thermodynamics in limiting what is physically possible in terms of economic activity. Daly was concerned with the size and durability of stocks of resources available on the planet, of which energy is the most important. The close relationship between energy and the growth economy is illustrated in Figure 5.4 which plots the size of the economy against the amount of energy each uses, for a range of the world's nations. This relationship makes it clear that, if we are to take seriously the energy limits of the planet, then we cannot continue to have a growth economy but must move instead to a steady-state economy. The SSE will be guided by certain principles to govern the use of resources:

1. For renewable resources, the rate of harvest should not exceed the rate of regeneration (sustainable yield);
2. The rates of waste generation from projects should not exceed the assimilative capacity of the environment (sustainable waste disposal); and
3. For non-renewable resources the depletion of the non-renewable resources should require comparable development of renewable substitutes for those resources.

(Costanza et al., 1997: 107)

This economy is respectful of nature's limits, hence Daly's suggestion, 40 years ago now, that 'the best use of resources would be to imitate the model that nature has furnished: a closed-loop system of material cycles powered by the sun'.

This is the picture of the SSE that is established by ecological economists: a system where we respect planetary limits and make policy in recognition of the fact that all resources are limited. We distinguish between those that are exhaustible or non-renewable and those that can be replenished. We ensure that our use of the exhaustible resources is especially frugal, that we minimize their use and recycle them. We are also careful in our stewardship of renewable resources, and our economic systems do not use them more rapidly than they can be replenished by natural systems. The economy as envisaged by ecological economics is thus no longer in conflict with natural systems but operates in balance with them.

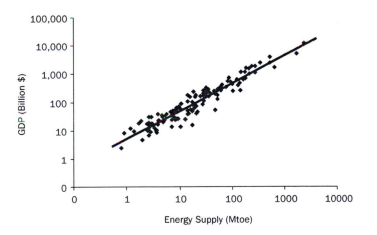

Figure 5.4 *Relationship between economic production and the supply of energy*

Source: Figure prepared by Daniel W. O'Neill using GDP data from the World Bank, 2008; UNDP 2007

Note: Mtoe = million tonnes of oil equivalent

It is apparent that the call for an SSE suggests a world with considerably lower levels of production – and hence consumption – than the prevailing culture demands. It is also notable that this theory grew up in the USA, which is the most intensively consumerist nation on earth. The vision that is portrayed in theory would have very serious implications for this level of consumption if it were to be introduced in practice. It is this conflict between the vision of a balanced economy and the reality of an economy based on a highly unequal distribution of resources that has inspired the contributions of the green and anti-capitalist economists whose work will be considered in the following two chapters. But before that, we will consider a case study that typifies the approach of ecological economics – the issue of pricing nature.

5.4. Case study: Pricing the priceless?

In the previous chapter, we saw how, when neoclassical economists enter the arena of environmental protection, they use their standard concepts of markets and prices. This case study focuses in detail on a paper that was written to provide a philosophical challenge to that thinking. It explores the idea that we might be able to measure the worth of a songbird, and was published in 1994. Funtowicz and Ravetz question the role of economics as what they refer to as a 'post-normal science', by which they mean that it cannot be enough to consider the consequences of economic behaviour within the existing paradigm, but it is necessary to ask questions about the ideological paradigm (in this case market economics) that is commonly taken for granted.

The specific feature of neoclassical economics that Funtowicz and Ravetz were challenging was the tendency to price every aspect of life. This has proved a contentious issue for ecological economists, with some leading proponents accepting the need for monetary valuation of ecosystems and their inhabitants, no matter how technically difficult this may be:

> So while ecosystem valuation is certainly difficult, one choice we do not have is whether or not to do it. Rather, the decisions we make, as a society, about ecosystems *imply* valuations. We can choose to make these valuations explicit or not; we can undertake them using the best available ecological science and understanding or not; we can do them with an explicit acknowledgement of the huge uncertainties involved or not; but as long as we are forced to make choices we are doing valuation.
>
> (Costanza et al., 1997: 143)

Funtowicz and Ravetz build on this argument by suggesting that, if we are to move to an economic paradigm that does not threaten the future of life on earth, we need to interpret 'value' more widely than would be the case in environmental economics – using its original, broad sense, rather than its market-based, narrow sense:

> In the first place, monetary price will be seen as a measure of one aspect of value reflecting one particular sort of interest, that which is mainly expressed through the commercial market . . . Some cultural goods are literally 'priceless', so that a people would rather die than give them up. A new enriched common language, which is not dominated by the worldview of one particular sort of stakeholder (expressed in the monetary standard), would come about when negotiators recognize the irreducible complexity of the issues at stake
> (Funtowicz and Ravetz, 1994: 198)

They identify three problems with the existing economic paradigm that the environmental crisis has highlighted:

- *The problem of uncertainties*. Ecological systems are so complex that we cannot have enough information to make decisions based on mathematical models. Rather, we need to rely on 'inherited, frequently unself-conscious craft skills'.
- *Management of quality*. Rather than consideration of quantity, we have to think in more subtle terms about the impacts of economic activity, and use judgement and interpretation rather than relying too heavily on scientific evidence.
- *Pluralism of methods and perspectives*. As the previous two points imply, mathematical modelling is insufficient for making decisions in such complex areas. Relying on experts to create such models is not sufficiently democratic, when all lives are at stake.

Funtowicz and Ravetz conclude that, 'The worth of a songbird definitely has its monetary aspect; but the endangered songbird is not thereby reduced to a commodity, any more than any other exemplification of love' (Funtowicz and Ravetz, 1994: 206). The very use of the word 'love' rather than 'utility' makes clear how far we have strayed from the scientific tendency of the neoclassical economist. It is clear that the sort of 'science' that the authors are arguing economics is capable of becoming – and needs to become – will be a far cry from the scientistic, intellectually self-confident and numerically secure discipline that dominates our universities today. Issues of uncertainty should be acknowledged and dealt with through the exercise of good judgement, rather than spurious mathematical modelling. Where judgements by experts are insufficient, public consultation and democratic processes of decision-making should be followed. The role of the expert is thus severely challenged. This case study reports on a paper that poses a challenging question, but one which the paper does not really answer. It leaves us with our own questions about how important a songbird, or a whale or an alder tree might be, and how much we would change about our economic behaviour to ensure that it did not become extinct. We cannot expect that such complex questions will find simple, numerical answers.

Summary questions

- Why is the second law of thermodynamics important to an ecological consideration of the economy?
- What do ecological economists mean by a closed system? And a closed loop?
- How is 'natural capital' distinct from 'human capital' and 'financial capital'?

Discussion questions

- Do you think it is possible to put a price on the continued existence of a species?
- What is the value, and what are the limitations, of applying an ecological perspective to the environment–economy tension?
- Can you 'think in systems'? Is it a useful exercise?

Further reading

Costanza, R., Cumberland, J., Daly, H., Goodland, R. and Norgaard, R. (1997), *An Introduction to Ecological Economics* (Boca Raton, FL: St Lucie Press): a wide-ranging, academic introduction to ecological economics from the US perspective.

Funtowicz, S. O. and Ravetz, J. R. (1994), 'The Worth of a Songbird: Ecological Economics as a Post-normal Science', *Ecological Economics*, 10: 197–207: I include this article since it offers the reader a thoughtful, philosophical consideration of what sort of economics might be appropriate to address questions relating to the environment.

Porritt, J. (2009), *Living Within our Means: Avoiding the Ultimate Recession* (London: Forum for the Future): Porritt's work is more general in tone, and most practical in orientation.

6 Green economics

We have taken the story as far as we can in terms of developments in the academy. Herman Daly's book *Steady-state Economics* was published in 1977, and gave a clear indication of the need to respect planetary and energy limits – and yet little has changed in the political or economic arena. The next two chapters cover schools of economic thought that take a more overtly political approach. Green and anti-capitalist economists would both subscribe to the view that there is an inherent political block preventing academic and theoretical economics from influencing the real economy to move in a direction that would be benign for the planet. Green economists identify themselves more with the tradition of political economy. In contrast to neoclassical and environmental economists, both of whom have faith in a perfect system of understanding that can generate beneficent outcomes, green economists argue that the environmental problems we face as a result of economic activity cannot be solved without fundamental political changes.

My own book on this subject summarizes the unique role of green economics as follows:

> Green economics has not grown up as an academic discipline but from the grassroots. It is distinct from environmental economics, which uses conventional economics but brings the environment into the equation; and ecological economics, which is still a measurement-based and academically focused discipline.
>
> (Cato, 2009: 206)

While clearly this is arguable, and the distinction between academic writing and grassroots activism is always a fuzzy one (as in my own case), nonetheless it is notable that the leading green economists whose work will be covered in this chapter tend to have spent their careers working for NGOs or political parties; this sets them apart from those economists whose work was covered in the previous chapter, who are much more frequently found in universities. That said, there are some ecological economists who are focused strongly on political economy, and the Marxist economists whose work is presented in the following chapter share this approach.

Section 6.1 outlines the key values on which a green economy would be based, and describes how green economists seek to re-embed the economy within the environment. Section 6.2 then proposes green economics as a new paradigm for economic life that is neither communism nor capitalism. Section 6.3 discusses some of the key policies needed to shift economic life towards a green economy. The following section, 6.4, provides practical examples of working green economies in the form of eco-villages and intentionally green communities. Finally, Section 6.5 offers a case study of complementary currencies as a fundamental part of an empowered local economy designed along green lines.

6.1. An economy with soul

Green economists accept many of the theoretical conclusions of the ecological economists, especially the importance of ending economic growth and developing a steady-state economy. Some of the central tenets of green economics are presented in Box 6.1.

Box 6.1

Ten design principles for a green economy

1. The primacy of use-value, intrinsic value and quality: the primary objective as the meeting of need rather than the generation of profit.
2. Following natural flows and working with the grain of nature, rather than engaging in a battle for domination of nature.
3. Waste equals food: the by-product from one production process should become an input to another production process.
4. Elegance and multifunctionality: the search for energy-efficient design and synergies in all economic processes.
5. Appropriate scale: rejecting the quest for economies of scale in favour of a size that is sustainable and just.
6. Diversity: seeking a range of forms of organization in place of the uniformity of the global marketplace.
7. Self-reliance, self-organization and self-design.
8. Participation and direct democracy.
9. Valuing and encouraging human creativity and development.
10. The strategic role of the built environment, the landscape and spatial design.

Source: Adapted from Milani (2000)

From a green point of view, it is difficult to separate the discipline of 'economics' from the other aspects of life, since the idea of holism – all aspects of life being interconnected – is central to a green philosophy. In terms of their view of the conventional economic model, green economists reject the concept of the **'externality'**. **Holism** suggests that everything is connected and that we need to address the planet as a system rather than looking at economic activity as separate from political activity, or industrial production as separate from healthcare. Clearly, if people working in sweatshops have unhealthy conditions, or the factories produce noxious pollution, this needs to be considered when making economic decisions, not just in terms of health policy. So the consequences of the pollution cannot be labelled as **externalities** if you take the system as a whole.

Conventional economics models the economy as a 'circular flow', with goods and services, labour and capital moving between firms and households. Green economists also need to consider the biosphere. When a conventional economist talks about pollution as an externality, we feel bound to ask: Where is that place outside our biosphere that the pollution goes to? From a green economics perspective the proverb, 'What goes around, comes around' – meaning that actions you take are likely to have consequences for you, even if these may be indirect and difficult to perceive – is more appropriate than talking about an external place where we can dump our wastes. The extended view of the productive economy as a circular flow including the biosphere is illustrated in Figure 6.1.

Figure 6.2 illustrates another important dimension of green economics: the significance of the positioning of the economy within social structures. The three circles on the left illustrate the orthodox economics view, with society, economy and environment as three separate circles that interact. The green economics perspective is illustrated on the right. The

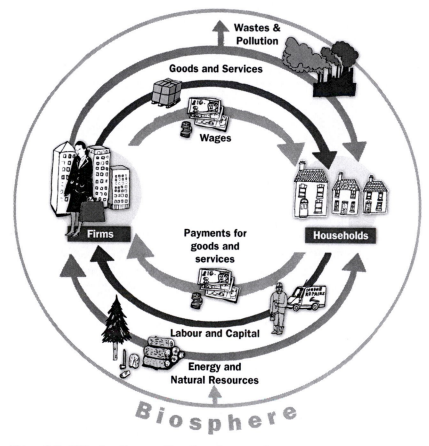

Figure 6.1 *Widening the consideration of economics beyond the classical economists' 'circular flow'*

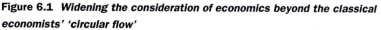

Source: Drawn by Imogen Shaw

economy is found within society, so economic decisions are illustrated as subject to social judgements and democratic decisions. Although this frequently happens within our existing economy – as with government regulation of the environmental health standards of cafés, for example – the market ideology suggests that the economy is separate from society. To a green economist, the embedding of the economy in social systems is a positive development: the market should be regulated to enhance social well-being, rather than maximizing profits and existing in some realm of supra-social pseudo-physical laws. Similarly, society needs to respect the fact that we are operating within a limited planetary system, and human culture needs to become re-embedded within the natural world.

Most of the debate explored thus far in the book has been conducted by the very narrow elite of Western, white men, who dominate the academic discipline of economics. It is a commitment of green economics that a broader range of perspectives should be included in the discussion, and so voices from the global South and women's perspectives are valued and encouraged. The movement away from the dry mathematics of the econometric method and towards more human, descriptive methods also helps to avoid the alienation of so many people from economic debate. Table 6.1 provides a simplified distinction between the 'hyper-expansionist'

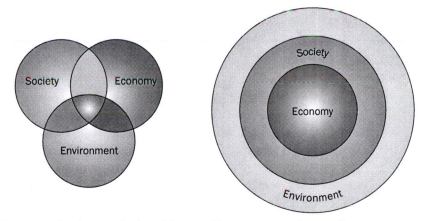

The conventional economic view of the interaction between economy, society and environment

The green economics paradigm: economy operates within social relationships and the whole of society is embedded within the natural world

Figure 6.2 *Rethinking the relationship between the economy, the environment and society*

Source: Figure 3.1 in Cato (2009)

Table 6.1 Comparison between the hyper-expansionist and sane, humane, ecological possible futures

HE	SHE
Quantitative values and goals	Qualitative values and goals
Economic growth	Human development
Organizational values and goals	Personal and interpersonal values and goals
Money values	Real needs and aspirations
Contractual relationships	Mutual exchange relationships
Intellectual, rational, detached	Intuitive, experiential, empathic
Masculine priorities	Feminine priorities
Specialization/helplessness	All-round competence
Technocracy/dependency	Self-reliance
Centralizing	Local
Urban	Country-wide
European	Planetary
Anthropocentric	Ecological

Source: Robertson (1985)

(HE) and 'sane, humane, ecological' (SHE) economies; it is important to note that the 'feminine' side actually reflects feminine values rather than suggesting in a simplistic way that the characteristics on the SHE side are the exclusive preserve of the female sex.

6.2. An alternative to capitalism that isn't communism

The following chapter will explore how a critique of capitalism as an economic system can be introduced into a school of environment-focused economics; however, I should draw attention to the fact that most green economists are opposed to capitalism, at least in the form we see it today, with the consolidation of businesses within many sectors on a global basis, and most economic power lying in the hands of corporate elites and beyond democratic decision-making. Porritt makes the case for a sympathetic view of capitalism and a pragmatic acceptance of its central role in our modern society:

> Like it or not (and the vast majority of people *do*), capitalism is now the only economic game in town. The drive to extend the reach of markets into every aspect of every economy is an irresistible force, and the benefits of today's globalization process still outweigh the costs – however substantive those costs may be, as we shall see. The

adaptability and inherent strengths of market-based, for-profit
economic systems have proved themselves time after time, and there
will be few reading this book who are not the direct beneficiaries of
those systems.

(Porritt, 2005: xlv)

However, the majority of green economists reject this pragmatic approach
and argue for an economic system that enables the economic
empowerment of local communities, and where cooperative forms of
ownership and management replace the hierarchical and profit-driven
businesses of the globalized economy. The extent to which this is a
capitalist economy is still a source of open discussion amongst green
economists, but most would support a diverse economy with a strong
emphasis on cooperatives, which enable people to share the value of their
labours in a way that is not possible within the capitalist business model.

As I know from personal experience, once you raise the issue of
capitalism as an economic form, the frequent response is that you must be
proposing some form of communism, as though there were only two ways
to organize economic life. While many green economists have benefited
from a Marxist analysis of the weakness of capitalism, this is not to
suggest that they are in favour of socialist solutions. To the extent that
these have been put into practice, greens raise a major criticism in terms
of the scale of economic organization and political control, and would
favour a focus on localization and devolving power over economic
decisions to the local community. Many socialist societies have replaced
the domination of the economy by private business with significant state
ownership of key economic sectors such as energy and banking. This is
anathema to a green economist, who would rather argue for small-scale
development and community control. (This is discussed further in
Chapter 12.)

As already mentioned, green economics critiques the male domination of
the neoclassical paradigm, and of economics as a university discipline;
the ecofeminists in particular have contributed a critique of the
masculinist perspective offered by neoclassical economics and the central
importance accorded to 'rational economic man':

> Economic man is fit, mobile, able-bodied, unencumbered by domestic
> or other responsibilities. The goods he consumes appear to him as
> finished products or services and disappear from his view on disposal
> or dismissal. He has no responsibility for the life-cycle of those goods
> or services any more than he questions the source of the air he
> breathes or the disposal of his excreta . . . Like Oscar Wilde's Dorian

Gray, economic man appears to exist in a smoothly functioning world, while the portrait in the attic represents his real social, biological and ecological condition.

(Mellor, 2006: 143)

The point is not merely to link patriarchal power relations in personal and political realms with a similar power imbalance in the economy, but also to draw attention to the way in which domestic and caring work, which has traditionally been carried out by women, is devalued in a market economy, especially when it is unpaid. Table 6.2 presents Mellor's comparison of the economic functions that are valued within a patriarchal economy and those – generally associated with feminine values – that are denigrated or neglected.

Hazel Henderson illustrated this selective blindness when considering what is 'economic' in her model of the global economy as an iced cake (illustrated in Figure 2.2). Economic decisions are taken based on a consideration of the top two layers (the public and private sectors), while the layers on which they depend – the work that people do reciprocally within communities and families, and the value nature provides, on which everything else depends – are not considered. The illustration is also important because it reminds us that the market has not always been central to our economic lives. Before about 200 years ago, communities provided for most of their own needs in terms of fuel, food and energy – trading was at the margin, and mainly involved surplus production. Greens today emphasize the importance of engaging in 'self-provisioning', both because it reduces the environmental impact of consumption and also because they foresee the breakdown of the complex and lengthy supply chains of the globalized economy and argue for the

Table 6.2 Valuation of activities and functions within the patriarchal economy

Highly valued	Low/no value
Economic 'man'	Women's work
Market value	Subsistence
Personal wealth	Social reciprocity
Labour/intellect	Body
Skills/tradeable knowledge	Feelings, emotions, wisdom
Able-bodied workers	Sick, needy, old, young
Exploitable resources	Eco-systems, wild nature
Unlimited growth, consumption	Sufficiency

Source: Mellor (2006)

importance of building resilience into local communities in the face of potential crises caused by climate change and the depletion of oil supplies.

Another defining feature of a green economy is its emphasis on post-materialist values, meaning that once people's needs for a reasonable standard of food and shelter are guaranteed, they have the space and time to consider more complex needs, including the need for a healthy society and flourishing environment. Thus, for a green economist, quality is more important than quantity – hence the importance of finding a more balanced measure of economic activity than gross domestic product (see Douthwaite (1992) and the further discussion in Chapter 9). Research from the London-based new economics foundation has investigated the contribution to happiness that is made by material as compared with non-material goods. Figure 6.3 illustrates the levels of life satisfaction of people by income and the extent of their social connections. It is clear that the very poorest people in society are unhappier than the rest of us, but people in the medium income group who have good social connections are happier than those in the highest income group, who are spending so much time earning money that they do not have time to maintain friendships. We might draw the conclusion that the economic activity we

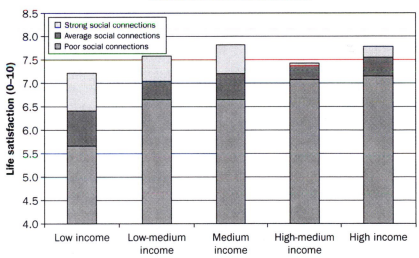

Figure 6.3 *Life satisfaction according to level of income and extent of social connections*

Source: Thanks to nef for permission to reproduce this graphic free of charge

are engaging in to earn money is no longer making us happy – and in fact could be leading us into isolated lives that cause us to neglect our emotional and spiritual needs.

6.3. Policies to create a green economy

Green economics is highly focused on policy, but since space is limited here I cannot do more than outline two specific policies and then discuss a green approach to economic development. The first and perhaps most important policy is the *citizen's income* (CI). This is a form of universal benefit that is paid to every citizen in a country, regardless of age and the extent of their involvement in the labour market. It is paid at a low rate that enables people to subsist, but most would be expected to engage in some form of paid work to supplement it. However, students or those who choose to pursue creative careers – such as artists or musicians – might choose to live on a very low income for some years while pursuing their education or creative development. It is important that entitlement is a birthright and does not depend on contributions paid, means-testing or availability for work. It could rather be seen as a national dividend. Because it does not rely on means-testing, the CI policy would remove the poverty trap, which prevents those on low incomes from taking paid employment because it would make them worse off if they lost all their benefits.

This policy is important because it provides a basis for the self-provisioning activities that can provide self-reliance, in the sense of being able to provide for one's needs largely and directly from one's own work (not *self-sufficiency*, in the sense of cutting adrift and trying to live entirely from one's own resources). Such a model of self-reliance also reduces dependence on an oil-hungry distribution system, but it raises important questions about land ownership and use. Those who own land are effectively enjoying the income from ownership of a part of the common wealth of the nation; those who do not should be compensated in the form of a national dividend that represents their share of this wealth. Conventional economists argue against the CI proposal, on the basis that it is a passport to paradise for the work-shy. A green economist might respond that there is no justice in a landowner living from **rental** income if a working person is not permitted to do the same. Another way of looking at this is that those without land can be thought of as effectively renting out their notional share of the land of the country. The CI could be considered to be their notional rent.

The question of land links to the other key green policy proposal that would shift power within our economy: a land value tax. At present, land ownership is highly concentrated. In the UK, for example, 64 per cent of the land belongs to 0.28 per cent of the population, and much of the land still belongs to descendants of the nobles who conquered England in 1066 (Cahill, 2001: 208). While green economists might criticize the ownership of our most valuable national resource in such an imbalanced and unjust way, conventional economists should also be critical of the inefficient use of it. For the classical economists, concentrated land ownership was a concern because the wealthy who live from rents have no incentive to use their land efficiently. At present, there may be an incentive to keep land unused while waiting for its value to rise – supermarkets do this with their 'land banks'. However, greens would not argue for more economic development, and the limited building that was considered to be necessary for the social good should be designed to achieve environmental and social goals, with the land tax operated in conjunction with the planning system.

The idea of a land tax was popularized in the late nineteenth century by Henry George, who built an international movement by arguing that the value gained from land should be shared between all members of the community. Land was marginalized in economic theory during the twentieth century, but green economists have made it central to their consideration of the economy. The central principle underlying the land tax is that the increase in land value arises from social investment – for example, the increase in the price of houses near the site of the Olympic developments in Greenwich, London – and therefore this increase in value should return to the community in tax rather than being kept by the individual:

> The arguments for a land-rent tax are to do with fairness and economic efficiency. Most of the reward from rising land values goes to those who own land, while most of the cost of the activities that create rising land values does not. This is because rising land values – for example in prosperous city centres or prime agricultural areas – are largely created by the activities of the community as a whole and by government regulations and subsidies, while the higher value of each particular site is enjoyed by its owner.
>
> (Robertson, 1999: 67–8)

Moving beyond the question of land alone, green economics has a particular view of how economic development might take place if it were to be truly sustainable. The central principle is that of the closed loop or closed system:

> Closed systems. It is here that the solution lies. And closed systems
> will take the form of local organisation, local economies. There will
> be no alternative. They will not be able to buy-in their needs, to import
> their way out of trouble. Local lean economies will not simply be a
> good idea; they will be the only option.
>
> (Fleming, 2004).

To explain in practical terms what this means, we might consider a market
town with its rural hinterland deciding democratically how to use the land
available to it. We can see already that economic planning would be based
much more locally, without the global supply chains we rely on today to
access our goods. So we would come to depend more on what we could
produce locally, and our wastes would also stay within this small unit.
Hence we would need to find practical uses for our waste products, for
example designing packaging that could biodegrade into useful soil to
produce crops for food or textiles. This takes us on to consider what green
communities might look like according to green economists.

6.4. Ecotopias in the here and now

As should have become clear by now, greens' conception of how society
should work is different from that of the prevailing orthodoxy. This is
partly about the global–local dimension that is discussed further in
Chapter 12, but it runs much deeper and broader than that. In my own
work, I have argued for a bioregional economy based on a new
consumption ethic. This involves a thorough re-exploration of what life is
for, and how we can live better lives with closer relationships in
functioning communities, rather than as the atomized, isolated consumers
that the globalized capitalist economy has created. Greens have been
inspired by the historical examples of communes to create eco-villages,
which can develop new lifestyles that minimize resource use, develop
social and technological innovations such as low-impact buildings and
community currencies (see more in the case-study in Section 6.5), and
model the low-carbon life that can ensure a safe future for humanity. The
Eco-Village Network is now a global movement that is experimenting
with sustainable community living.

A concept that helps to illustrate how a green economic life might offer a
higher standard of living with a lower level of material wealth is
'conviviality'. Here is how Ivan Illich (1974) describes the concept:

> I choose the term 'conviviality' to designate the opposite of industrial
> productivity. I intend it to mean autonomous and creative intercourse

among persons, and the intercourse of persons with their environment;
and this in contrast with the conditioned response of persons to the
demands made upon them by others, and by a man-made environment.
I consider conviviality to be individual freedom realized in personal
interdependence and, as such, an intrinsic ethical value. I believe that,
in any society, as conviviality is reduced below a certain level, no
amount of industrial productivity can effectively satisfy the needs it
creates among society's members.

(Illich, 1974: 11)

Conviviality requires a rethinking of what an economy is for. It means a
reduction in working hours and more time spent in relationship and
problem-solving in communities; it means more make-do-and-mend and
less monetary exchange, and local identity rather than brand loyalty.

Richard Douthwaite argues for the development of a 'peasant economy',
by which he means an economy where 'most families own their means of
making their livelihoods, be this a workshop, a fishing boat, a retail
business, a professional practice or a farm' (Douthwaite, 1996: 32). He
contrasts this with the industrial economy that dominated the twentieth
century:

The difference between the industrial and peasant systems is not only
that one seeks to minimise the returns to labour and maximise those to
capital, while the other wants to minimise the return to borrowed
capital and maximise a wide range of benefits including income for
the group involved. There is also a difference of scale. An investor-
owned, industrial-system venture can grow extremely large through
mergers or by ploughing back its profits, the techniques which General
Motors – with 251,130 people on its payroll and an income which
exceeded the **GNP** of all but twenty-one countries – used to become
the biggest company in the world in terms of employment at the
beginning of the 1990s. Peasant projects, by contrast, tend to stay
fairly small.

(Douthwaite, 1996: 32)

Greens believe that, in economic terms, small is beautiful because it
allows people to show more respect and responsibility for their part of the
world. Perhaps not-in-my-backyard-(NIMBY)-ism is not such a bad
thing: if everybody took responsibility for their own backyard, since each
part of the planet is somebody's backyard, the whole planet could be
safeguarded. Similarly, if we were required to take responsibility for our
waste within our local economy rather than exporting it, we would be
likely to ensure that it was minimized; and if we had access only to the
resources that our local area provided, we would use them more carefully.

To summarize:

> From a radical green perspective, reduced affluence, self-sufficiency, small-scale living, localized economies, participatory democracy and alternative technologies – all are key ingredients of an ecologically benign and socially just society.
>
> (Pepper, 2010: 42)

Less radical greens are concerned that this strategy of building self-sufficient communities may be a form of escapism that is either a distraction from the need for political change, or a preserve of the wealthy. North (2009) considers that localization offers new ideas about 'livelihood' as opposed to 'employment' that are inherently anti-capitalist, but is critical that the smaller economies proposed by radical greens may be impractical and inefficient, especially in terms of economies of scale. They may also be authoritarian and inward-looking. Pepper also finds the proposed return to an era of local production and consumption to be impractical, in an era when communities are much more fluid than they once were, and both products and production processes far more complex and specialized:

> They cannot (realistically) reverse what has been a hugely strong historical (even ahistorical) drive towards functional differentiation – spatially and between economic, political and technological subsystems – creating a highly mobile and pluralist world society where traditional bonds are irrevocably loosened.
>
> (Pepper, 1993: 227)

For some greens, the response to such a critique would be to draw attention to the enormous consumption of fossil fuels that has enabled this sort of society and economy, and the carbon dioxide emissions associated with it that the atmosphere can no longer assimilate. The other fork of this argument is that identified in the subtitle to Douthwaite's book *Short Circuit: Strengthening Local Economies for Security in an Unstable World* (1996): it may not be a question of choice. As the triple crunch of financial crisis, ecological crisis and peak oil intensifies, we may simply not be able to rely on the lengthy supply chains and international trade negotiations that have underpinned the provision of our basic needs for the past 30 years or so.

6.5. Case study: Complementary currencies

There is a dominant strand of scepticism about money amongst green economists, who argue that the way money is created as debt by banks has led to both inequality and environmental pressure (Douthwaite, 1999). Since money is lent into circulation rather than spent into circulation, those with lower incomes are obliged to borrow it from those with larger incomes – and pay for the privilege. The debt that supports the creation of credit has to be repaid – and with interest – so that those who incur this debt have to work producing goods that can be sold so that they can earn wages to pay their debts. This production entails the use of materials and energy, and thus generates the pressure to grow that is the central driving-force of a capitalist economy. The majority of green economists argue for monetary reform so that money becomes once again a source of 'common wealth' (Robertson and Huber, 2000; Mellor, 2010).

Money also tends to leak out of local economies into elevated global circuits where it is used simply to generate more money, not facilitating the production and exchange of real economic activity. To build the strong local economies that green economists believe are necessary to reduce our planetary impact, we need to find ways to stop value leaking out of the local economy. The new economics foundation (nef) has developed the concept of the 'leaky bucket' to illustrate how money that comes into a local economy leaves almost immediately if it is spent in a local branch of a chain. If it is spent in a local shop, by contrast, it circulates more times – a process referred to as the 'local multiplier' – thus building strength into the local economy.

One way of ensuring that money cannot leave the local economy is by transforming it into a local currency. By definition, this can be spent only in local shops and is issued as a sort of voucher by a community group. An example is the Berkshare, which has been operating in the US state of Masschusetts since autumn 2006. The notes are accepted by more than 365 businesses, and more than one million were circulated in the first nine months of the scheme. According to the scheme website (http://www.berkshares.org), 'The people who choose to use the currency make a conscious commitment to buy local first. They are taking personal responsibility for the health and well-being of their community by laying the foundation of a truly vibrant, thriving local economy.'

The Transition movement in the UK has followed this lead, with four communities – Totnes, Lewes, Stroud and Brixton – setting up currencies in 2008 and 2009. All are exchanged one-for-one with sterling so that consumers and businesses can rely on the credibility of the national currency – although the aim is that the money should be spent many times in the local economy before being swapped back. A survey undertaken a year after the launch of the Lewes pound indicated that traders who had joined the scheme were positive about it, with 75 per cent saying that it offers an opportunity to support the local economy. However, a weakness is that the currency often circulates only once before being switched back to sterling. Stroud has introduced two features to help to counteract this: a 'demurrage' charge, which means that a small percentage has to be paid at regular intervals to keep the currency valid, and a redemption fee for businesses who seek to swap their Stroud pounds back to sterling. For more details of how community currencies work, see North (2010).

Summary questions

- What would a green economist make of the concept of an 'externality'?
- What does a green economist mean by a 'closed-loop economy', and why is this important?
- Why do we still have economic growth even if it is no longer making us any happier?

Discussion questions

- Would a green economist argue that women should be paid for housework?
- What is the difference between employment and a livelihood?
- What is the use of a local currency if everything sold in the local economy is produced in China?

Further reading

Cato, M. S. (2008), *Green Economics: An Introduction to Theory, Policy and Practice* (London: Earthscan): my own introduction to the field, which is accessible to a general reader.

Douthwaite, R. (1992), *The Growth Illusion: How Economic Growth Enriched the Few, Impoverished the Many and Endangered the Planet* (Totnes: Green Books): a seminal work that raises many of the questions of concern to a green economist.

Robertson, J. (1989), *Future Wealth: New Economics for the 21st Century* (London: Cassell): a radical view of economics from 'the grandfather of green economics'.

7 Anti-capitalist economics

We could roughly characterize the discussion in the chapters so far as representing a movement along a continuum from a scientistic and market-focused economics towards a more intuitive and people-centred approach. This chapter is somewhat different. It describes the way that economists in a distinct tradition – broadly following the critical political economy of Karl Marx – have brought the environment into their worldview. It also includes others who, while they have been influenced by Marxist thinking, would not be comfortable being identified as part of this tradition. The fact that both these contributions have been brought together in one chapter is largely a question of convenience, and does not imply that they share a worldview. What the writers included in this chapter share is a view that the source of the environmental problem is to be found in the structure of the globalized capitalist economy, and particularly in the issue of who owns and controls its powerful economic organizations. Thus we will find a repeated emphasis on the excessive power of corporations and the impact this power has on the inequitable distribution of resources and the democratic deficit that characterizes both economic and political relations in the twenty-first century.

For many economists in the Marxist tradition, the twentieth century started with great enthusiasm, but ended in disappointment. They recognize that the dominant ideology of the twenty-first century will be that of sustainability, in some form or other, and are convinced of the importance of ensuring that this issue is inextricably linked with the focus on ownership and control of economic life that is the heart of a Marxist critique:

> Just as socialism can only hope to remain a radical and benign
> pressure for social change by assuming an ecological dimension, so
> the ecological concern will remain largely ineffective . . . if it is not
> associated in a very integral way with many traditional socialist

demands, such as assaulting the global stranglehold of multinational capital.

(Soper, 1996: 82)

This chapter begins by exploring how economists who are in the Marxist tradition have woven the environmental concern into a Marxist analysis, focusing especially on what has been called the 'second contradiction of capitalism'. Section 7.2 discusses the contribution of economists who have for the most part been significantly influenced by the Marxist critique, and how they have sought to extend Marx's analysis of economic power to describe a world dominated by global corporations. The contributions in this section can be considered as anti-capitalist, although they are not explicitly Marxist. As discussed in Section 7.3, other critiques of capitalism have responded by seeking ways to re-include the interests of the excluded and disempowered – specifically women and the poor of the global South – in the economic debate. Finally, Section 7.4 outlines the proposal for participatory economics as a socially just way to organize economic life.

7.1. Capitalism, nature, socialism

The heading for this section is taken from the title of a key journal, which has provided a fertile test-bed for the germination of a socialist solution to the environmental crisis. Anti-capitalist economists have been influenced by Marx's writings and have concluded that '[t]he evidence favours the judgement that capitalism is not ecologically sustainable' (O'Connor, 1998: 236).

Marx considered the natural limit to the productive process to be an inevitable problem for capitalism:

> Capitalist production has not yet succeeded and never will succeed in mastering these (organic) processes in the same way as it has mastered purely mechanical or inorganic chemical processes. Raw materials such as skins, etc., and other animal products become dearer partly because the insipid law of rent increases the value of these products as civilizations advance. As far as coal and metal (wood) are concerned, they become more difficult as mines are exhausted.

(Benton, 1996: 68)

Marx was similarly critical of the exploitative and environmentally destructive impact of capitalistic agriculture. However, like all economists, Marx was a product of his time, and his concern for the environment was limited to specific ecosystems and spaces, since the global economy had not, in the late nineteenth century when he was writing, reached such a scale as to threaten the whole global eco-system. Hence, while 'Marx and Engels had interesting (and damning) things to say about the effects of capitalism on the productivity of soils and forests, slum housing, urban pollution, the destructive mental and physical effects of certain types of concrete labour' (O'Connor, 1998: 329), they failed to take seriously enough the limited nature of natural resources and the second law of thermodynamics (Pepper, 1993). Therefore, much of the work summarized in this chapter extrapolates and extends Marxist theory beyond Marx's own writings.

For Marx, capitalism is a system that generates and thrives on conflict and crisis. In his own work, the central conflict is between the owning and working classes, and the crisis arises from the allocation of productive value, as profit extraction leaves an ever-smaller share to be distributed amongst those who work, earn and therefore have the spending power to buy goods. Once we introduce the concept of a limited planet into this framework we see that the same concepts remain useful, but undergo a change of emphasis:

> An ecological Marxist account of capitalism as a crisis-driven system focuses on the way that the combined power of capitalist production relations and productive forces self-destruct by impairing or destroying rather than reproducing their own conditions . . . Such an account stresses the process of exploitation of labor and self-expanding capital, state regulation of the provision or regulation of production conditions, and social struggles organized around capital's use and abuse of these conditions.
>
> (O'Connor, 1998: 165)

According to Marx, every commodity within a capitalist economy has a use value and an exchange value. Exchange value is measured in terms of other commodities or in terms of money, as the universal source of 'value', whereas use value is the inherent value of the commodity either for immediate consumption or as an input to a further production process. Capitalists generate profits by selling the products of their factories for a price greater than that of their use value, but the workers who make the products are paid only the use value as wages. Thus 'surplus value' can be extracted as profit.

Marx identified a contradiction inherent within the capitalist system arising from the inability of the productive forces to generate sufficient surplus value to pay for profits and large enough incomes to buy the products of economic activity. A proportion of the value of production stays with workers, whereas another portion is accumulated as capital by owners. So there is always less value being spent in the economy than there are goods being produced. This would lead inevitably to insufficient demand for the products of economic activity. The latter is a crisis of overproduction or under-consumption, which is central to the 'first contradiction of capitalism'. The distinctly 'ecological' dimension to the Marxist analysis is what O'Connor (1988) terms the 'second contradiction of capitalism', in which capitalism expands to such an extent that it undermines its 'productive conditions', degrading the environment and exhausting the inputs that it needs to make products and create profits by selling them.

Marx distinguished three types of conditions of production: 'personal conditions', the human contribution in terms of labour and creativity; 'natural or external conditions', the environment and natural resources; and 'general communal conditions', the communities and spaces where the labourers live. 'Sustainable capitalism would require all three conditions of production to be available at the right time and right place and in the right quantities, and at the right fictitious prices' (O'Connor, 1988: 243). The environmental crisis not only demonstrates the limitation on all three 'conditions' but also the inevitability of a continuous rise in their prices, thus undermining profitability, and therefore contributing to the inherently crisis-prone structural character of capitalism.

O'Connor concludes by proposing a version of 'ecological socialism', which accepts the central tents of Marxist economics but extends these to take into account the planetary limit:

> I use the term 'ecological socialism' to distinguish theories and movements that seek to subordinate exchange value to use value and abstract labor to concrete labor, that is, to organize production for need (including the self-developmental needs of workers), not for profit. Ecosocialism, thus defined, problematizes both the capitalist labor process and also the structure of use value and needs (consumption). In this sense, ecosocialism seeks to make traditional socialism live up to its own critical ideals.
>
> (O'Connor, 1998: 331)

Box 7.1 summarizes O'Connor's account of some comparisons between traditional socialism and ecological socialism.

Box 7.1

Comparisons and contrasts between traditional socialism and ecological socialism

Traditional socialism	*Ecological socialism*
Universalistic, quantitative critique of capitalism/exchange value (effective demand, liquidity, etc.)	Particularistic, qualitative critique of capitalism/use value (integrity of the site, specific tasks in the labour process, the individual, etc.)
The extraction of some of the value of labour as problematic	Labour that results in physical production can also be problematic
Focus on production and circulation of capital (workplace and markets)	Focus on conditions of production (society and the state)
Creation of inequality between classes and continents	Degradation of productive forces
North's economic debt to the South	North's ecological debt to the South
Presupposes availability of inputs such as land and labour	Availability and types of inputs (e.g. renewable energy) seen as problematic
Nationalization of the means of production	The means of production to be owned by communities/co-operatives
Downgrades the issue of land/community	Upgrades the issue of land/community
Uneasy balance between top-down economic planning and worker control of industry	Tension resolved via top-down planning and user control of industry via the participatory democratic state
Economic struggles to redistribute wealth and income fought at national level	Economic struggles to redistribute wealth and income fought at international level

Source: Summarized and simplified from O'Connor (1998: 334)

Some ecological economists have been influenced by Marx's work on equity and his identification of the problematic economic consequences of 'concentrated ownership and control' (Costanza et al., 1997: 35). They identify unequal ownership as the fundamental engine of planetary exploitation, as in the example of the destruction of the Amazon rainforest, which is driven by the poor's need for land to farm and by the ability of the rich to pressurize political authorities to subsidize their cattle-ranching. However, according to Sarkar (1999), the work of ecological economists such as Herman Daly suffers from internal contradictions because they are wedded to a capitalist worldview and find it hard to conceive of an economic organization that is not based on capitalist principles. He suggests that a Marxist analysis would strengthen their line of attack on the existing economy, particularly in terms of prescriptions for change.

The inability of capitalism as an economic system to function when beset by its own contradictions is the critical message of Marxist economics. In the environmental sphere, this is expressed as the inherent unsustainability of a system that must grow if it is to survive. However, other concepts from Marx's writing are also applied to the environmental crisis. For example, 'alienation' (Marx's concept of the divorcing of 'man' from his essential cooperative and creative nature through exploitative capitalist work relations) is extended to cover the holding in thrall of contemporary populations to market society – 'subverting community by atomizing individuals into selfish globules of desire' (O'Connor, 1998: 328). Just as their work is effectively outside the workers' control, likewise it is argued that their consumption is controlled by a dominant ideology created by advertising and status competition. Likewise, Marx's concept of 'commodity fetishism' (the perception of goods as having a mystical force and power within society) is discussed with reference to the 'affluenza' of unsatisfying but addictive consumption that typifies contemporary Western society (Soper, 1996).

7.2. Reincluding the excluded

Marx wrote during the heyday of capitalism, when individual capitalists could be clearly identified, and were often self-made men who had risen from their own proletarian origins to build vast business empires. Today, the situation with regard to ownership is much murkier, with amorphous corporations holding vast amounts of power in the global economy through their CEOs and board members. Even more indirectly, it is frequently argued that we all have a stake in the capitalist economy through our investments in pension funds and even in the firms we work for through employee share-ownership schemes. Anti-capitalist economists would argue that this has changed the nature of owners but done nothing to change the nature of the ownership structure itself. If the profit system relies on the extraction of surplus value, how is that achieved in today's global economy?

The anti-capitalist critique of neoliberal globalization has focused on the domination by corporations (Korten, 1995) and their ability to exercise more political power than national governments and supranational organizations. The past 50 years have been characterized by the abdication of a political role by national governments and the increasing liberalization of economies (especially the finance sector), allowing corporate domination of resources, distribution, labour markets and finance. Critics such as Korten (1995) and Kovel (2002) identify this excessive power exercised by profit-driven corporations as a key cause of the environmental crisis, because governments who might protect the environment on behalf of citizens have given away their power to do so. Their prescription is that governments should take back this power, and exercise it for the benefit of citizens. They also propose strengthened local economies and a reduction in global trade (a theme that is developed further in Chapter 12). Kovel in particular has been overtly critical of green parties for not being explicitly anti-capitalist in their economic policies.[9] His own eco-socialist position identifies the nature of operation of a capitalist economy as the fundamental explanation for both ecological crisis and societal collapse. Kovel's work helps to define the distinction between green economics and eco-socialism: his view is that the self-sufficient community proposed by green localizers is a fantasy, and that the destructive forces of a capitalist economy must be confronted and defeated before any sustainable economy can be built.

If an anti-capitalist ecological critique of global capitalism focuses on exploitation of certain oppressed groups, its proponents also argue for the empowerment of these groups – both within the discourse of economics and in the economy itself. Ecofeminists have drawn particular attention to the way in which the contribution of women and the world's poor is ignored in economic theory, even though they are in reality the base of any functioning economy. To illustrate this point, Maria Mies produced her 'iceberg model' of the global economy reproduced as Figure 7.1:

> The layers making up that enormous invisible economy might be considered as built up in an order of increasing monetisation, with contracted and wage-labour exposed above the line . . . Under the waterline are the 'colonies'. Both internal and external to a national economy, colonies may be defined by race, by gender, and by nationalism in the distribution of resources. They include nature. They

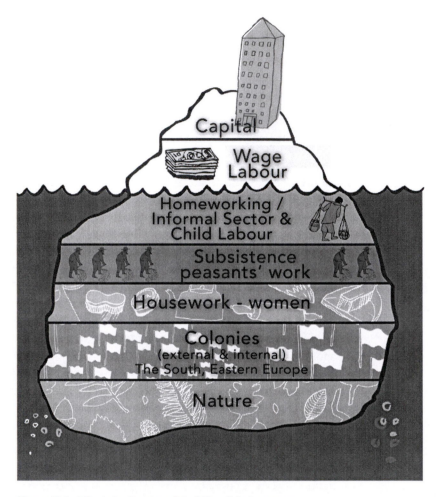

Figure 7.1 *Mies's iceberg model of the global economy*

Source: Figure 5.1 in Cato and Kennett (1999); redrawn by Imogen Shaw

exemplify the continuum of relationships based on violence, often
with military oversight, which serve to extract resources for the benefit
of powers outside the colony.

(Mies, 1999: 49)

Ecofeminist critics in the socialist tradition have extended the analysis of
the 'mode of production' (the way in which productive forces such as
land and labour are combined within a particular social and economic
system) to include the 'mode of social reproduction', arguing that without
the contribution of those who care for children, and the other necessary

reproductive labour in the 'domestic' sphere, there would be no workers to produce goods for sale in the market. This invisible work has been undertaken predominantly by women. Mary Mellor identifies a link between the hidden nature of this 'marginalized work' and 'enforced caring labour', which relates to the physical needs of human beings, and the 'disembedding' of the economy from nature. The traditional economic account foregrounds the rational and the mental, dislocating human beings from their frail and vulnerable bodies – much as we have lost contact with our place in the natural world. She develops this argument in terms of the 'rational economic man' who is considered the typical 'economic agent' within neoclassical economic theory:

> 'Economic Man' is not young or old, sick or unhappy, 'he' does not have pressing domestic demands that cannot be ignored or put off. As a result, the artificially boundaried set of human activities that is called the 'economy' fails to acknowledge its true resource base and the way it is parasitical upon sustaining systems of unpaid social labour and the natural world. As a result these are exploited and damaged.
> (Hutchison et al., 2002: 158)

Neoclassical economics simplifies what is in reality a complex web of relationships that constitute an economy. In doing this, it overlooks the important perspectives of women, indigenous people and the planet herself (Hawthorne, 2009), including the interests of, and our responsibilities towards, future generations. In response, ecofeminists have developed the 'subsistence perspective', where subsistence means the provision of one's basic needs directly by one's own labour:

> This concept was first developed to analyse the hidden, unpaid or poorly paid work of housewives, subsistence peasants and small producers in the so-called informal sector, particularly in the South, as the underpinning and foundation of capitalist patriarchy's model of unlimited growth of goods and money. Subsistence work as life-producing and life-preserving work . . . was and is a necessary precondition for survival; and the bulk of this work is done by women.
> (Mies and Shiva, 1993: 297–8)

Some critics have suggested that the term 'subsistence' is a patronizing and Western-centric one, and that the term 'sufficiency perspective' should be used instead, and should include in its definition the assumption of a higher level of well-being than mere survival.

On the one hand, this concept draws attention to the vastly unequal share of resources that women and the poor of the world receive through the allocative processes of the globalized market economy. Women are the poorest of the poor globally within the context of the increasing 'feminization of poverty'. But on the other hand, this 'sufficiency perspective' also indicates that these oppressed groups are actually providing a model for the eco-sufficient life that we will all need to adopt if we are to live in balance with nature and that might provide an alternative epistemological basis for an ecological economics (Salleh, 2009). The question of the level of life that is sufficient is still an open one amongst ecofeminists and green economists (see also the discussion in Section 10.4). To some extent, the question cannot be answered until we first draw a boundary around the areas of the world from which we can legitimately draw resources, which is discussed in the bioregional perspective outlined in Section 12.3.

Political economists based in the South have also provided perspectives that have traditionally been marginalized within the economics discipline. Martin Khor (2001a), from the Third World Network in Penang, offered a trenchant critique of the orthodox neoliberal patterns of economic globalization that were being proposed as a solution to global poverty. He outlines three types of 'liberalization' that have had a negative impact on economic prospects for the nations in the South: finance, trade and investment. In the case of finance, the freeing up of global money markets has led to a vast increase in speculative movements of money, with only 2 per cent of international exchange transactions now relating to the

exchange of goods and services, as opposed to purely financial transactions. Investment has similarly resulted in more value being extracted from than anchored to the benefit of indigenous people. Meanwhile, other evidence makes clear that trade has benefited only a small number of the larger countries of the South (Nayyar, 1997), and has been to the advantage of only a small number of economic players – what Marxists would term the capitalist class in the global South – in those countries; and this advantage has been bought at a great environmental and social cost (Lines, 2008). In response, some countries are pioneering a practical implementation of the ideas of eco-sufficiency, as in the 'sufficiency economy' model adopted by Thailand (UNDP, 2007).

7.3. Changing the world

While Marxist economists are critical of capitalism, they nevertheless acknowledge its dynamic and adaptive characteristics – it may be dominated by crisis and conflict, but these in themselves drive a process of technological and social innovation that will enable a creative response in the face of ecological crisis. This 'creative destruction', as Joseph Schumpeter termed it, is a constitutive feature of capitalism's protean and adaptive capacity. Sarkar (1999) proposes a form of 'market socialism', which follows nature's way in including a diversity of forms of economic organization:

> Private enterprise would be allowed in market socialism . . . State enterprises would, of course, dominate the commanding heights of the economy, such as the banking and credit system. But between the two poles there would be various other kinds of ownership: co-operatives, joint ventures between public and private sectors, joint stock companies, companies fully owned by their workers, companies largely owned by their workers, and so on.
>
> (Sarkar, 1999: 184–5)

He proposes social ownership of the means of production, which eco-socialists favour not just because it denies private owners the opportunity to expropriate workers' surplus value and profit, and is therefore inherently just, but also because it enables socially beneficial decisions to be reached to tackle the environmental crisis. It also has the effect of recreating and sustaining solidarity and community between people, production and place. A profit-driven system might prioritize other objectives that are not in the general interest. An example of the prioritization of profit-driven over social ends is the investment of research funding into the development of slimming medicines at the expense of cures for fatal diseases such as malaria, because the latter diseases affect mainly the poor. However, Sarkar is keen to point out that social ownership of the means of production does not mean state ownership; rather, ownership by communities or workers would be encouraged, as in employee or consumer co-operatives, leading to an economy that reflected the wishes of communities because they had genuine ownership of productive organizations.

The importance of taking control over productive forces, and especially of maintaining the value of production for the producers themselves via the establishment of worker co-operatives, is seen as key to ensuring social justice, as well as reducing economic pressure on the environment. Milani (2000: 183) proposes a balance between small-scale, empowered local economies and the reshaping of the market economy by 'conscious regulation through a democratized state'. He also underlines the importance of using a participatory process to realign the economy, the need for a strong regulatory structure controlling production, and a form of 'green municipalism' that provides strong social protection and public services (see also the conclusion in Pepper, 1993).

Box 7.2

Windpower as our mutual friend

The Recession that blighted the developed economies during 2008–10 resulted in calls for a **Green New Deal**, which received considerable rhetorical support from policy-makers. Yet while politicians stated the need for an urgent transition to a sustainable economy powered by renewable energy, in July 2009, the Danish company Vestas decided to close the only factory in the UK that was producing wind turbines. The employees of the factory took over the premises and rapidly joined the RMT union. The union leader, Bob Crow, asked with understandable frustration why money could be found to keep the UK's banks afloat when the much smaller sum that would be needed to keep this industry of the future in Britain was denied. For many anti-capitalists, the answer was simple: the interests of capital dominate those of labour and of the planet. Anti-capitalist economists might use this as evidence of the incompatibility of a sustainable economy with capitalism when the profit motive is the dominant driving force of a capitalist system.

But should the solution be a return to centralized state control over economic life, so that important decisions about the nature of electricity generation could be made by central political authority? Many eco-socialists have looked enviously at the rapid progress towards a sufficiency economy that Cuba made following the end of cheap oil imports from the Soviet Union;[10] and

China seems best placed to shift its economy rapidly towards a low-carbon future precisely because it does not have to worry about selling these changes to a sceptical electorate. It is significant that the Chinese Communist Party has made the creation of what it calls 'ecological civilization' central to its economic and political vision for China in the twenty-first century. Others on the left argue that the answer is not to sign away our hard-won right to power over our lives, or to return to the days of public ownership and central planning; it is rather to call for ownership and control at the local level. Vestas offers a perfect example of how a mutual future would achieve the advantages of rapid change without the political opposition that arises when people feel they are powerless pawns in a game that is played for the benefit of others.

Vestas cited a low level of demand for its turbines by the UK market – a result of planning delays – and a reluctance to invest the necessary money to adapt the factory to the sort of turbines that are favoured by UK wind power developers. Planning consent has been difficult to obtain in the UK, since local communities who will bear the brunt of wind farm development gain nothing in return. If the turbines were community-owned, such opposition would be much less intense (Cato et al., 2008). The other side of the coin might be the buyout of the firm by its own workers, creating a co-operative on the production as well as the consumption side. Thus an eco-socialist economy could rely on government to create a supportive framework, supporting communal ownership and worker buyouts through advantageous planning guidance or direct financial support, and allow positive relationships to develop between producers and consumers of environmentally sensitive products and services.

Source: http://gaianeconomics.blogspot.com

Mention should also be made of socialist critics of globalized capitalism who share the opinion that it is the key cause of ecological crisis but are opposed to a strong state or suprastate authority as the solution. Foremost amongst these is Murray Bookchin, a former anarchist and founder of the social ecology movement, who proposed community engagement in developing small-scale participatory communities as the most sustainable form of social and economic organization. Bookchin's pioneering work linked the social imbalances created by capitalism with the parallel ecological imbalances as early as the 1960s (see more detail in Section 2.6). In a discussion that prefigured much of the debate amongst contemporary green economists, Bookchin argued for small-scale, self-reliant communities, producing most of their own food and using local renewable forms of energy.

Here we can see much in common with the green proposals for future economic life, with a greater emphasis on local and community ownership and control, but essentially the same vision of ecological units smaller than today's urban centres, which are self-reliant but not self-sufficient, which maintain the value of their resources and labour, and which are governed through participatory democratic systems. So there are commonalities, but anti-capitalist economists are more likely to be sceptical of new market forms, such as fair-trade brands and farmers' markets, which can easily become co-opted by the mainstream market economy – so that Tesco becomes the major retailer of fair-trade coffee, and Asda redesigns its stores to include striped awnings and locally produced vegetables.

7.4. Case study: Participatory economics and the World Social Forum

As we have seen throughout this chapter, the critique of capitalism is not new and has been acquiring a growing number of adherents in recent years. What is more problematic and contested is the prescription for the ideal society to replace the globalized capitalist market. Non-capitalist societies in the recent past have been dominated by both state ownership of productive forces and centralized planning. A more empowered and localized alternative is the participatory economics or 'parecon' proposed by Michael Albert and Robin Hahnel (Albert, 2003). The central proposals of the system are:

1. Economic decisions made by a combination of workers' and consumers' councils, with people's influence on decision-making proportional to their commitment to the enterprise.
2. Financial remuneration in an enterprise is made depending on effort rather than on bargaining power, so that working longer hours or in harsher conditions results in being paid more. Effort is assessed in personal terms, so that having better technology or better physical ability does not result in achieving a higher return.
3. Rather than a division of labour based on skill or power, tedious and unpopular work will be shared and will be mixed with high-skilled, more personally satisfying work.
4. Markets to be replaced by participatory planning.

Michael Albert describes the benefits of the system as follows:

> Parecon doesn't reduce productivity but instead provides adequate and proper incentives to work to the level people desire to consume. It doesn't bias toward longer hours but allows free choice of work versus

leisure. It doesn't pursue what is most profitable regardless of impact on workers, ecology, and even consumers, but it reorients output toward what is truly beneficial in light of full social and environmental costs and benefits.

(http://www.zcommunications.org/life-after-capitalism-and-now-too-by-michael-albert, accessed 13 September 2010)

Supporters of parecon suggest that a version is being developed in Latin America, with left-wing presidents pursuing policies to share resources with the poorest in their societies, for example the governments of Hugo Chavez in Venezuela and Evo Morales in Bolivia. Perhaps the most impressive example of participatory economic planning is the city of Porto Alegre in Brazil, venue for the first World Social Forum in 2001, which was established as a rhetorical response to the elitist World Economics Forum held annually in Davos, Switzerland.

Since 1989 spending on public services in Porto Alegre has been decided by a participatory budgeting process. A process of decentralized and democratic decision-making in the city's neighbourhoods identifies spending priorities for the $200 million. annual budget for infrastructure and services. The process is not fully participatory – about 50,000 of the city's 1.5 million inhabitants take part – but is distinct from the systems of representative democracy that decide budgeting in cities of a similar size the world over.

Summary questions

- What do eco-socialists mean by 'the second crisis of capitalism'?
- What do the ecofeminists mean by 're-embedding' in the environment? Do you find it a useful concept?
- Can we manage the transition towards a green economy without the incentive of the profit motive?

Discussion questions

- How does the nature of ownership relate to the environmental crisis? How might co-operatives help to solve the crisis?
- Is there any reason why, within an ecologically sensitive economy where the means of production are socially owned, women would have more economic power?

- Do you think a democratic state, a centrally planned state, or a system of participatory planning would be more effective at introducing strict environmental controls on economic activities?

Further reading

Foster, J. B. (2002), 'Capitalism and Ecology: The Nature of the Contradiction', *Monthly Review*, Sept. Available online: http://www.monthlyreview.org/0902foster.htm: a brief introduction to the thesis of a 'second contradiction of capitalism'.

Kovel, J. (2002), *The Enemy of Nature: The End of Capitalism or the End of the World* (London: Zed): a polemical work that nicely encapsulates the political nature of the work covered in this chapter.

Mellor, M. (2006), 'Ecofeminist Political Economy', *International Journal of Green Economics*, 1(1–2): 139–50: a brief and valuable introduction to the ecofeminist perspective.

Pepper, D. (1993), *Eco-socialism: From Deep Ecology to Social Justice* (London: Routledge): a widely read early contribution to the task of linking Marxist and ecological thinking.

Part III
Issues and policies

8 A range of policy approaches

Having established that a certain environmental threat is real, it becomes the task of policy-makers (politicians and their civil servants) to devise policies to address the threat. They have a range of policies at their disposal. We can broadly divide these into two types: regulatory measures (examples include outright bans, the introduction of quotas, rules about labelling), where government uses its political power to impose limits and requirements on economic actors; and incentive-based measures – including taxation or the introduction of permits. In this chapter, we consider the advantages and disadvantages of some of these policies in a theoretical perspective.

Before reaching the comparison of available policies, Section 8.1 reflects on the policy discourse by discussing what has become known as the 'ecological modernization' debate. Section 8.2 compares the role that regulation and incentive-based instruments have played in controlling the economy's impact on the environment. Section 8.3 then considers the problem of measurement, which is key to assessing the success of policies, and is especially problematic in the area of environment-related policy-making. Section 8.4 addresses the issue of changing behaviour to reduce the need for policy interventions, before the final section, 8.5, outlines one policy that includes some aspects of all the above policy types as a case study – the End-of-Life Vehicles Directive of the European Union.

8.1. How much change and who should make it?

For many who attempt to devise policies to tackle environmental problems, there is no essential tension between the way our economy and society function and a healthy environment, and therefore there is no need for fundamental social or economic change. This is known as 'ecological modernization': 'The basic tenet of ecological modernization is that the zero-sum character of environment–economic trade-offs is more apparent than real' (Barry, 1999: 191). In other words, we can solve the environmental crisis without making any significant structural changes to the way our economy is organized. This might be read as an argument for 'green capitalism', where economic structures such as the corporation and the market are maintained, but the products they produce change so that they are produced in a less energy-intensive way, perhaps:

> While the process of modernisation of the economy (capitalist industrialism) has caused environmental problems, the solution to them lies in the direction of more or better modernisation, not, as the early green movement and many radical environmental groups still hold, in radically altering or indeed rejecting modernisation. That is, what is required to cope with contemporary and future environmental problems is a suitably ecologically enlightened or rational evolution of modernisation; that is, 'ecological modernisation'.
>
> (Barry, 1999: 192)

The counter-position is that the change we need to address, the environmental challenge, is deep and fundamental. If capitalism is to survive at all – and as we have seen, this is not a given for many environmentally focused economists – then it must undergo significant change from its present form (Porritt, 2005). This is the debate that underlies all debates about policy, although it is often implicit, with the world as it is being taken as a given by politicians and their civil servants.

The other question that must be addressed before policies can be implemented is who the prime actors should be in taking the steps necessary to tackle the negative impact of the global economy on the environment. The different schools of economic thought outlined in Part I of this book take different positions on this question, which fundamentally revolve around the extent of the role that the state should play. Neoclassical and environmental economists see a limited role for the state, which should merely remove examples of 'market failure' to enable the market to acheive efficient outcomes. For example, in the case of pollution control, the government's role would be confined to deciding the limit on the pollutant (based on objective scientific advice) and then issuing rights to pollute, which companies could trade between themselves. Finding ways to reduce the emission of pollutants would become a competitive, profit-driven process, thus stimulating innovation and technological development.

Ecological economists are more sceptical about the ability of the market to solve environmental problems, offering as evidence the fact that the global free-market economy has resulted in the pressing environmental crises that we see in many areas of life. They suggest a greater role for government in terms of banning certain harmful processes absolutely. Green economists also suggest a strengthened role for political authorities, although they are likely to suggest more devolution of power to devise regulatory systems to local authorities, operating within global frameworks. They share with the anti-capitalist economists a deep scepticism about the ability of corporations to tackle environmental problems that they consider to be primarily a result of those corporations' profit-driven activities. The anti-capitalist economists suggest the need for major state intervention, including ownership of some of the most potentially polluting industries, such as power generation or construction. When we face crisis, they might argue, as in the case of the financial crisis that hit the world economy in autumn 2008, we realize that in reality the state is the economic actor of last resort. The environmental crisis is more fundamental than the financial crisis, and therefore requires energetic and widespread state action.

Table 8.1 attempts to summarize the two key variables that determine what policies should be followed: whether the necessary changes are considered to be superficial or structural; and whether a market or political solution is favoured. If we begin with the thinking of the ecological modernizers, we see that a state solution might result in firm

Table 8.1 *Who should make the policy change, and how deep should it run?*

Nature of problem	State	Market
Superficial	Regulatory state	Managerialism and environmental targets
Structural	Social ownership and subsidiarity	Environmental empathy rather than profit as business motive

regulation of polluters and those who over-exploit resources. A more market-focused solution would be what we see in the developed Western economies today: corporations offering to manage the environmental crisis through the production of eco-friendly products, and improved environmental performance and reporting.

For those who consider that the workings of global and national economies need to be fundamentally restructured if we are to avert environmental catastrophe, a market solution might be possible only if businesses change their core motivation so that ecological and social objectives are as important as the profit motive in determining their strategies. Critics of ecological modernization are more likely to favour a strong role for the state, and frequently also the genuine social ownership of potentially harmful production processes. It may be that the appropriate level for ownership is that of the local communities who will potentially suffer the harmful effects of the productive processes, rather than the nation-state (see the longer discussion of globalization vs. localization in Chapter 12).

8.2. Regulation or incentive-based instruments?

Regulation describes a policy situation where the state uses its democratic power to legislate to prohibit productive processes that result in negative environmental consequences. It dominated initial responses to environmental problems, although market solutions have become more fashionable. To some extent, this has been the result of deliberate lobbying by their proponents, who include neoclassical and environmental economists, and who frequently refer to regulation as 'command and control' – an inherently pejorative phrase, which conjures images of powerful and intrusive (and probably communist) states. In reality, however, well-enforced legislation has experienced considerable success historically – as in the example of the British Clean Air Acts, which were passed in the 1950s following large numbers of deaths caused by urban pollution – and, when clearly drafted and strictly enforced, it can help to correct market failures in private capitalist economies. An important benefit of this type of direct control of potentially damaging economic behaviour is that it can be imposed in cases of uncertainty that do not lend themselves to market-based instruments, since the costs that need to be charged are not easy to assess. Regulatory controls also tend to prevent emissions, which is usually cheaper than cleaning them up after they have occurred – no matter who eventually pays that bill – so they are inherently more efficient.

In addition to outright bans on certain types of destructive behaviour, governments can introduce limits on the amount of a certain pollutant that companies can emit – generally known as a quota. They can also create a regulatory regime within a certain market sector, such as introducing labelling to indicate that a food product has been produced without genetically modified (GM) ingredients, or that a wood product has been made using timber from a sustainably managed forest.

Several problems emerge in the literature on regulation as a method to tackle environmental problems. First, what is referred to as 'imperfect information', that is to say that it is difficult for policy-makers to define what a safe level of a pollutant might be and therefore to set levels for quotas. In some cases, such as nuclear waste, the consequences of the pollution may be unpredictable or difficult to measure – or may not express themselves for many generations. Second, any regulatory framework is subject to lobbying pressure from the industries whose behaviour will be limited by the legislation, and this may extend to those industries funding scientific research to influence the regulatory authority's view about the potential risk from their activities, as happened in the case of climate change. Figure 8.1 provides a schematic illustration of the relationship between the various players who make up the regulatory framework, and indicates the complex and intertwined nature of their interactions. Although policy is democratically determined in theory, the route of influence via voting (indicated by letter A in the diagram) can be counteracted by lobbying activity on the part of business interests (indicated by letter B in the diagram). Third, regulation is extremely expensive to enforce, with large numbers of inspectors needed to check regularly that polluters are sticking to the law, especially since there is no positive incentive for polluters to do so – a problem that is addressed directly by market-based measures.

The pro-market philosophy that dominates modern economic theorizing portrays the market as a solution to most problems, and (notwithstanding

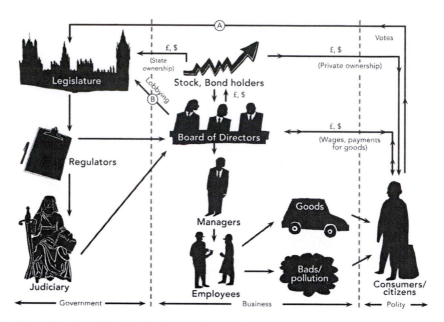

Figure 8.1 *Graphical illustration of the complex relationships between government, firms and citizens*

Source: Author's graphic based on Figure 8.1 from Kolstad, 2000

the fact that most environmental problems are defined as 'market failure' of one form or another) market-type solutions in the form of financial incentives to change corporate or individual behaviour are currently much in vogue. In a culture where we have been trained to respond to price signals, some of these sorts of instruments can work well, and they can be an efficient means of achieving efficient allocations – for example, the sale of permits to produce a certain pollutant from a factory that can easily switch to less polluting technology to one that cannot. In this way the factories that can reduce pollution most easily are the ones that do so. The other main type of incentive-based instrument is an eco-tax, which may be imposed on the pollutant itself (as in the Scandinavian fertilizer taxes) or on the product, which is itself environmentally damaging (as in the proposed carbon tax on fossil fuels). An idea of the range of policies based on incentives available to policy-makers is given in Box 8.1.

Box 8.1

Types of incentive-based instruments for environmental management

- Taxes on pollution emissions (Pigouvian taxes or charges)
- Product charges (levied on products whose use causes environmental damage, such as CFCs, carbon-based fuels, agricultural chemicals and fertilizers)
- Subsidies for pollution abatement (similar to taxes in concept but not in distributional consequences), especially for agriculture and sewage treatment
- Marketable permits for pollution emissions
- Creation of property rights for open access and other environmental resources
- Creation of economic incentives for acting in the common interest

Source: Costanza et al. (1997)

The theory behind these market-based instruments is the neoclassical economics outlined in Chapter 3; its specific application in this case is illustrated in Figure 8.2. The curves are labelled as MDC, for marginal damage cost, i.e. the extra damage to the environment that is caused as production increases; and MAC, for marginal abatement cost, or the cost of reducing pollution (there is more discussion of what these marginal curves mean in Chapter 11). The point where these curves intersect indicates the most efficient rate to set the tax to affect behaviour, in this case at T*, where the level of pollution is EP*. This is the point at which reducing pollution any more would cost more than it would save; similarly, if there were more pollution emitted, the cost of restoring the environment would cost more than it would to not emit the pollution in the first place.

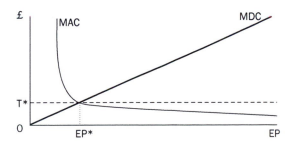

Figure 8.2 *Setting the optimal rate of a tax on pollution*

Source: Drawn by Imogen Shaw based on a figure in Pearce and Barbier, 2000, p. 201.

This is how a tax would be established in theory. In practice, creating just the right level of incentive is difficult, since policy-makers do not know the costs of abatement or damage, and polluters have an incentive to be economical with the truth. The London congestion charge (see Box 8.2) is an example of a tax that provided incentives for Londoners to walk, cycle or use public transport. It was designed after consultation with the relevant stakeholders and – although introduced in the face of strenuous opposition – has now become accepted and achieved its policy objectives.

Box 8.2

The London congestion charge

The congestion charge in London was motivated more by irritation at the slow pace of traffic in the city than by environmental concern, but it has nonetheless provided an important example of how traffic can be reduced in one of the world's largest cities. By the 1990s, traffic was moving more slowly in the UK's capital than it had been at the beginning of the twentieth century – before cars had been invented! Following his election as Mayor in 2000, Ken Livingstone launched an 18-month period of public consultation, one outcome of which was a decision to launch a congestion charge based on area licensing rather than parking levies. Considerable research and modelling was undertaken to predict the correct level of the charge to deter the desired number of people (30 per cent) from continuing to drive into the capital. In February 2003, a daily charge of £5 was introduced between 7am and 6.30pm on weekdays; this was increased to £8 in July 2005. Research predicted that, at a rate of £5, car miles travelled in central London would be reduced by 20–25 per cent, and total vehicle miles would be reduced by 10–15 per cent.

Table 8.2 Impact of the congestion charge on traffic in London

Type of vehicle	% change
Cars	−34
Vans	−5
Trucks	−7
Taxis	+22
Buses	+21
Motorcycles	+6
Bicycles	+28
All vehicles	−12

Car traffic was reduced by 34 per cent, representing up to 70,000 journeys no longer made by car on a daily basis. Details of changes in road-traffic journeys are given in Table 8.2. Transport for London estimates that about half these journeys are now made by public transport; a quarter divert to avoid the zone; 10 per cent have shifted to other forms of private transport including bicycles; and a further 10 per cent have either stopped travelling or changed their time of travel. There have been sharp rises in journeys by bus, taxi and bicycle. Meanwhile, travel speeds have increased by some 17 per cent. The reduction in vehicle usage within the charging zone was greater than expected, leading to less revenue than predicted. Since the revenue is dedicated to improving public transport services this is slightly problematic. The London congestion charge appears to have been a political and environmental success, and has encouraged changes in behaviour towards less polluting forms of transport, thereby reducing CO_2 emissions. It is also an example of a tax that is flexible, since the rate can be increased or decreased depending on the relative balance of traffic and public transport desired by the city's residents.

Source: Leape (2006); Cato (2008): Ch. 10

Except for those of us who are, for ideological reasons, particularly wedded to one form of policy or another, the logical response is to work with a mix of the available options, and to choose policies that work well in a given situation. Regulatory methods can be burdensome and expensive to administer, but in an area such as nuclear pollution, where unregulated emissions can be deadly to large human and animal populations, regulation is clearly the best response (Cropper and Oates, 1992). On the other hand, businesses are comfortable with policies based around economic incentives, which make them more likely to be rapidly successful in tackling other environmental threats, such as emissions of toxic gases. Figure 8.3 illustrates how different policy measures might be appropriate for different types of ecological damage. Where measurable damage is minimal, property rights approaches may well be sufficient to tackle the problem – low-level effluent from a factory might be an example of such a low-grade pollutant. Once we have measurable damage and reduced productivity, we need to introduce financial incentives – here we might be thinking about significant air pollution from factories. Then as we move into the area of non-sustainable, long-term damage – such as climate-changing gas emissions, or radionuclides – we need to introduce regulatory legislation. Of course, while this is fine in theory and a useful way to guide our thinking, if the world were as simple as indicated in this figure, we would be unlikely to be suffering the environmental crisis that we see all around us.

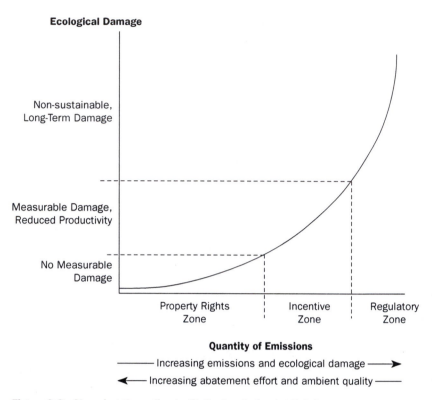

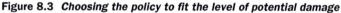

Figure 8.3 *Choosing the policy to fit the level of potential damage*

Source: Drawn by Imogen Shaw based on a figure in Cumberland, 1994

8.3. Measurement issues

Once policy-makers have intervened to affect economic activity, they need to have some way of measuring whether their interventions are effective. This is particularly difficult in the area of environmental protection. On the one hand, ecological systems are hugely complex and interrelated, with numerous interactions and feedbacks. We can see this complexity most clearly in the example of climate change, where the range of predictions as to how any particular change might impact on the climate system is vast. On the other hand, we have an economy made up of people who are, if anything, even more unpredictable. When the Irish government introduced a 10c tax on plastic bags they expected to raise money to spend on environmental projects; however, people stopped using plastic bags almost entirely. So designing economic policy to protect the environment is a challenging game.

Before policies are introduced, we need to have an idea of how we will measure their success. Economic policy-making in most nation-states is dominated by one measure: economic growth, usually calculated as gross domestic product (GDP). This is taken as the dominant measure of human progress – so long as there is more economic activity, questions about whether it is benign or destructive have rarely been asked. We will consider this in more detail in Chapter 9, which looks at the issue of economic growth in detail, but it is worth mentioning here because the discussion of policy needs to be set within a framework where economic growth is a measure of success accepted without question by conventional economists. This has a powerful constraining impact on the range of policy prescriptions that are countenanced.

As we have seen in Chapters 3 and 4, economics as most commonly practised by political advisers and researchers is dominated by mathematical methods, so quantity prevails over quality in assessing the performance of an economy. The method of cost–benefit analysis (CBA, see Section 3.3) is a demonstration of this focus on measurement and costing in monetary terms, which is criticized by environmentalists and ecological economists for its inability to include some of the most valuable and important aspects of life. This leads to morally questionable techniques such as the costing of human lives (see Box 8.3) and equally difficult problems with measuring outcomes in terms of happiness, an attempt to impose numerical values on intangible emotional states that was pioneered by the philosophical school of utilitarianism.

Box. 8.3

The cost of life on earth

Critics of CBA have two main grounds of attack. The first is that there is something inherently unethical about attempting to put a price on human life, especially when it emerges that lives enjoyed in the richer parts of the world have a higher value than those spent in poorer countries (see Section 4.1). The second objection is about the wild assumptions that need to made in order to calculate the probabilities on which CBA is based, and the even wilder conclusions in terms of the cost of lives that these give rise to. Both flaws are demonstrated clearly in the work of Richard Posner, who has attempted to calculate the cost of the extinction of the entire human race, which he set at US$600 trillion. He also confidently assigned probabilities to entirely unpredictable natural disasters: 70 per cent in the case of a climate-change-related loss of US$1 trillion in 2024, for example.

On the basis of this supposed risk, Posner makes suggestions as to how much money should be invested in policies to attempt to avert climate change (Geertz, 2005).

Posner has responded to his critics by suggesting that what they suggest is an attempt to put a price on human life is, in reality, merely a 'mathematical transformation':

> Suppose that it is discovered by studies of people's behavior that the average person would be willing to incur a maximum cost of $1 to avoid the one-in-a-million chance of being killed by some hazard that a proposed project would eliminate. And suppose that 2 million persons are at risk from this hazard and that the proposed project (which for simplicity I will assume has no other benefits) will cost $3 million. Since each of the persons benefited (in an expected sense) by the policy would pay only $1 to avoid the hazard, for a total of $2 million, the benefits are less than the costs. An equivalent way of putting this is that the life-saving project can be expected to save the lives of only two people, each of whom 'values his life' at 'only' $1 million ($1/.000001), and so the total benefits are only $2 million and are less than the costs. As I said, this is just an arithmetical transformation of an analysis that values risks rather than lives.

> (Posner, 2000: 1160)

The desire to arrive at fixed values that can form the basis of policy-making is understandable, and was the motivation behind the development of CBA. However, such a technique should always be used with caution, and with clear caveats placed on the results it generates. The spread of this method beyond its moral boundary, and into areas where no data exist on which to base calculations, offers only a highly unreliable basis for policy-making.

8.4. Cultural and behavioural change

The discussion thus far has been conducted in the language of the conventional policy-maker – we might imagine him in a smart suit and tie. It is time now to loosen the collar, and maybe put on some sandals, as we veer off into some more fundamental approaches to tackling environmental problems. Climate change in particular is an issue that requires us to think seriously about how we live. The changes that we will need to make to our lives in response to both climate change and peak oil pose deep questions about our quality of life – perhaps even about the meaning of life. Many of the decisions we feel bound to take in our working lives – for example, to maximize the profits of the firm we work for by outsourcing production to China – cause problems that threaten us in our roles as responsible citizens or parents.

To environmental philosophers, our ethical approach to the environment should be the seat of change. How would we approach the environment if, rather than seeing it as something outside the window, we felt that we were truly a part of it? We would not then need to introduce policies to prevent companies from polluting because their employees would intrinsically resist poisoning their own air. If we were truly embedded in our environment, then protecting it would become a personal commitment rather than something governments needed to devise policy to force us to do. This is the approach of the environmental philosophers who develop the concept of 'ecological citizenship'.

Ecological citizenship suggests that strict policy-making might militate against our best environmental behaviour. Frey and Jegen (2001) analysed a mass of international research into the effect that compensating people financially for acts and services they performed freely had on their willingness to continue in their virtuous behaviour. They distinguished between intrinsic and extrinsic motivations, i.e. those things that people do because they are ethically motivated or just feel good about them, rather than those things they do because there is some material incentive or disincentive. In the case of volunteering, they found that paying volunteers actually reduced their amount of volunteering. In a similar vein, an Israeli childcare centre that began charging parents who picked up their children late found that more were doing so – the explanation is that, once it became a financial matter, parents felt they had an implicit contract; whereas when they felt guilty about their lateness they tried much harder to be on time.

The example of research most relevant to the theme of this book concerns the willingness of communities to accept the siting of a noxious facility in their backyard, in this case a nuclear waste repository in Switzerland. Conventional economists would suggest that the solution is to provide financial compensation. Initially, 50.8 per cent of residents agreed to have the repository in their community, with 44.9 per cent in opposition. The next stage was to offer variable rates of compensation to local residents, who were then surveyed again:

> The respondents were asked the same questions, whether they were willing to accept the construction of a nuclear waste repository, but it was added that the Swiss parliament had decided on a substantial compensation for all residents of the host community. While 50.8% of the respondents agreed to accept the nuclear waste repository without compensation, *the level of acceptance dropped to 24.6%* when compensation was offered. The amount of compensation has no

significant effect on the level of acceptance. About one quarter of the respondents even seem to reject the facility simply because financial compensation is attached to it.

(Frey and Jegen, 2001: 603–4)

These findings have important implications for policy in this area, since we may conclude that taxation or charging may be expensive in itself but may also be inefficient if it discourages people from undertaking environmentally friendly behaviour that they might have engaged in anyway without the incentive. And in other areas they might now look for incentives before changing their behaviour (Berglund and Matti, 2006). Policy-makers should beware of crowding out people's natural motivations to do good, and to respect each other and their environment. The danger with applying tools such as CBA is that they assume a selfish motivation that may not exist. However, assuming such selfishness can create a self-fulfilling prophecy, and can train citizens to be self-serving and less virtuous.

So if policy-makers might do more harm than good with their restrictive policies, how might they encourage the sorts of shifts in behaviour and moral (spiritual, even?) outlook that would enable a protective ethic towards the planet? Dobson (2003) suggests the rewarding and celebration of pro-environment behaviour – perhaps designating environmental 'champions', or rewarding particularly well-embedded citizens through the award of medals. Rather than struggling to live within our carbon quota, perhaps we might find ourselves receiving the Order of the Lapwing or being designated Green Man of the Year?

8.5. Case study: The EU End-of-Life Vehicles Directive

In 1997, the EU began developing a directive that would change the nature of the relationship between producers and waste, and challenge the reliance on built-in obsolescence to guarantee future markets for producers. The approach began in the automotive sector, with the End-of-Life Vehicles (ELV) Directive, which was adopted by the European Parliament in September 2000 (information is taken from Defra, EU and BERR websites). The automotive sector is a huge part of most developed economies, a significant user of raw materials and energy, and a producer of waste. UK Government data show that currently:

- Around 2 million vehicles are scrapped in the UK every year.
- Around 1.2 million of these go to vehicle dismantlers in the first instance.

- The remaining 0.6 million go directly to scrap yards.

According to the Directive, the car continues to belong to the manufacturer throughout its life, and the manufacturer is responsible for its safe disposal once it is no longer functioning. The objectives of the Directive are to increase the proportion of components in cars that are recyclable and recycled, and to encourage car manufacturers to design new cars that are more durable and easier to recycle. From 2007, EU citizens have been able to dispose of their vehicles free of charge. Articles 5 and 7 of the Directive require that:

- Owners must be able to have their complete ELVs accepted by collection systems free of charge, even when they have a negative value.
- Producers (vehicle manufacturers or professional importers) must pay 'all or a significant part' of the costs of take back and treatment for complete ELVs.
- Rising targets for re-use, recycling and recovery must be achieved by economic operators.

When we consider this Directive in terms of the discussion in this chapter, we can see that it has aspects of all three of the policy types that were distinguished. It is a piece of legislation that is regulating the market, and manufacturers who do not follow it, for example by fly-tipping vehicles, could be fined or imprisoned. However, it also demonstrates aspects of incentive-based policy, since manufacturers will save money if they design their cars so that they will be cheaper to dispose of in future. In addition, consumers and especially producers are being encouraged to change their behaviour and their consumption patterns, by thinking more about the whole life cycle of the products they buy, rather than considering a car to be an item for which they have only a limited responsibility.

The ELV Directive may be the shape of policy to come. The Waste Electrical and Electronic Equipment (WEEE) Directive has been in force in the EU since August 2004. Its aim is to reduce the dumping of electronic goods, which frequently contain hazardous materials, and again to encourage producers to manufacture such goods with more reusable components. However, producer behaviour has proved hard to change and in 2008 the EU revised the regulations to cope with the fact that only one-third of the materials covered by the Directive were being recycled – the rest were being illegally dumped or exported outside the EU.

Summary questions

- What are the advantages of a regulatory policy? And what are its costs?
- What problems do you anticipate as a result of market-based policies to tackle pollution?
- What policy would you suggest to reduce the speed at which we are using up the limited supplies of copper in the earth's crust?

Discussion questions

- Can we make environmental policy without putting a price on human life?
- Do you think introducing a congestion charge in your local community would be beneficial for the environment?
- Which would make you more likely to recycle your waste: being given a medal, being paid to do it, or being fined for not recycling?

Further reading

Barry, J. (1999), *Rethinking Green Politics: Nature, Virtue and Progress* (London: Sage), ch. 10: offers an introduction to the concept of 'ecological modernization'.

Dobson, A. (2003), *Citizenship and the Environment* (Oxford: Oxford University Press): a philosophical discussion of how we might behave as ecologically responsible citizens, and what might encourage us to do so.

Kolstad, C. D. (2000), *Environmental Economics* (Oxford: Oxford University Press): includes a great deal of useful information on the advantages and disadvantages of different policies to address environmental problems.

9 Economic growth

For thousands of years, the human race existed in a fairly stable relationship with the earth. There are renowned examples of populations expanding beyond the resources available in the niche they were inhabiting (Sale, 2006a), and of warfare between small-scale human populations over local resources. However, the situation facing the human race now is of a different order of magnitude, because the size of the human population, its rate of increase and its level of consumption mean that we have run up against the limits of all resources available on the earth. This, at least in the view of many environmentalists, is the cause of many of the prevailing social and economic crises. Why human population began to increase so rapidly is a question that has preoccupied economists with a concern for the environment. The response to this population expansion, and the increasingly sophisticated lifestyle and consumption pattern that have accompanied it, has been a constant and accelerating rate of growth in economic activity. As shown in Figure 9.1, following a long period of a more-or-less steady level of population, in the past 200 years both population and economic activity have mushroomed, and this growth increased markedly following the discovery and exploitation of fossil fuels.

Growth is a basic and unavoidable feature of the global economy as it is presently structured. Politicians target economic growth as the key measure of the success of their management (or lack of management) of the economy. This chapter explores the relationship between economic growth and the state of the environment. We begin by considering the positive, even eulogistic, view of economic growth that is shared by the classical and neoclassical economists. Section 9.2 takes a more critical perspective, presenting the views of those economists who have questioned the consequences of an economy that expands infinitely within a limited ecosystem. We then take a detour, in Section 9.3, to examine exactly what is meant by growth, and how useful it is as a measure of what our economy is achieving. Section 9.4 explores the concerns of

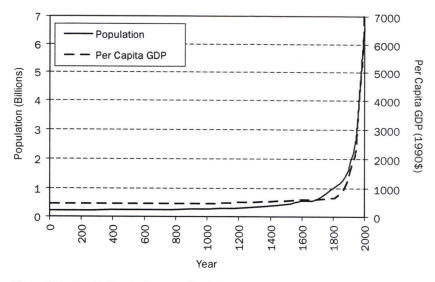

Figure 9.1 *Population and per capita GDP, 1–2006 AD*

Source: Figure prepared by Daniel W. O'Neill using historical population and GDP data in Maddison (2008)

economic commentators who see a connection between economic growth and inequality, and who question how the recognition of planetary limits might affect our view of social justice. In Section 9.5, we reach what must surely be the most significant question. Why are economists so concerned with the quantity of economic growth rather than the quality of the life that we are leading as a result of it? Finally, in Section 9.6, I present a case-study of the one measure of economic well-being rather than economic output.

9.1. Growth is good

It seems a fairly uncontroversial statement that orthodox economists favour economic growth, which is equated with progress, advancement, higher consumption rates and a better quality of life. In the words of Adam Smith, 'The progressive state is in reality the cheerful and the hearty state to all the different orders of society. The stationary is dull; the declining melancholy.' However, it is worth noting that Smith and the other classical economists – Thomas Malthus and David Ricardo – were pessimistic about the prospect for growth in the long run, and the reason they gave was the limitation of what they saw as the basic resource: land. The classical economists' view was that, in the long run, population growth and diminishing returns would ensure that all surplus value created through production was transferred to rents. This would mean an eventual end to profit and a corresponding end to economic growth (Daly and Townsend, 1993).

This was also the view of J. S. Mill, who conceived of economic progress as 'a race between technical change and diminishing returns in agriculture' (Pearce and Turner, 1989: 7). His conclusion was optimistic: technological progress would allow us to meet all of our material needs and live in a state of economically stationary comfort, pursuing intellectual, artistic and spiritual interests. However, such a paradise was to arise in what he defined as a 'stationary state': 'It is perhaps telling that over a century earlier John Stuart Mill had advocated a "stationary state" economy based around workers' co-operatives' (Dresner, 2002: 105). This view was shared by Marx, who theorized that perpetual growth was impossible, and drew social and political conclusions about what the end of growth would mean in terms of conflict over the value of production between those who owned the plant and the capital to control it, and those who had only their labour to sell.

As we saw in Chapter 3, however, the neoclassical economists' view is that innovation and technology can enable the infinitely more efficient use of resources – a view that is sometimes referred to as 'Promethean', after the Greek myth of Prometheus, who stole the ability to make fire from the Gods and was horribly punished for his arrogance. The aim of the neoclassical model is to achieve a lasting and steady rate of economic growth through managing the factors of production: capital, labour and technology. In extensions of the theory that rely on the work of Schumpeter, enterprise is sometimes included as a fourth factor of production. We can note that land no longer features in this theory, and that the emphasis has shifted away from the accumulation of capital and towards the acquisition of more and more complex technology. The theory includes the standard assumption of neoclassical economics that an economy tends towards equilibrium, which will be achieved when the economy has reached a point where the ability of capital to generate further growth is exhausted. This is defined as the 'steady state', which causes confusion with Herman Daly's quite different conception of a 'steady-state economy', which we will arrive at shortly.

The general agreement amongst neoclassical economists is that neither pollution nor resource limitations should operate as constraints on perpetual economic growth. However, Homer-Dixon (1994) has suggested that scarcity of material goods and resources could impede economic efficiency, since groups within society might operate to improve their sectional interest, making the harnessing of human creativity more difficult and thus reducing economic output. He has also suggested that human ingenuity might be diverted towards dealing with problems of scarcity management (he does not go so far as to extend this to the potential environmental crises), and that this might lower the future economic growth path of an economy. A conventional economic view will thus permit that:

> There is some evidence that in many poor economies depletion and degradation of natural resources – such as agricultural land, forests, fresh water and fisheries – may be a contributing factor in social processes that destabilize the institutional and economic conditions necessary for innovation and growth.
>
> (Pearce and Barbier, 2000: 40)

However, the solution, they say, is not to restrain economic growth but rather to ensure the efficient management of scarce resources.

For neoclassical economists, the desirability of growth is unproblematic: 'On the whole, endogenous growth theorists have not been concerned with the contribution of natural resources to growth or with the role of innovation in overcoming resource scarcities' (ibid.: 37). They think it is the nature of economies to grow, and the role of economists to ensure that they do this in a steady and balanced way. Neoclassical economics includes a great deal of theorizing about economic growth – indeed, Robert Solow was awarded the Swedish Bank Prize (sometimes referred to as the Nobel Prize for economics) for developing the endogenous growth theory – but includes very little consideration of how a continuously growing economy might impact on the environment.

9.2. Systems thinking and the steady-state economy

The *Limits to Growth* report (1972) was significant for introducing important concepts from systems thinking to the discussion around the impact of the economy on the environment (its authors' specific contribution is covered in Section 2.2). One of these key concepts is that of a feedback: a feedback occurs when the output from a process has an impact on that same process in the future: feedbacks can be positive (reinforcing) or negative (balancing). An example of a negative feedback loop would be the relationship between population and mortality (the relationship first identified by Malthus): the relationship is mediated by food shortage so that as population rises, demand for food increases, leading to food shortages, starvation and death, thus naturally bringing population numbers back to their original level. However, in planetary terms, some of the feedback loops that are most threatening in terms of the future of human societies are positive, as in many examples of concern to climate scientists. To give just one example, as the polar ice melts due to rising temperatures, so the reflective capacity of the planet diminishes, and more of the sun's heat is absorbed rather than reflected back into space, leading to further warming. Clearly, this is a 'positive' feedback in a technical sense only, because its consequences for life on earth are far from positive. Some of the most important relationships between economic and demographic variables are illustrated in Figure 9.2.

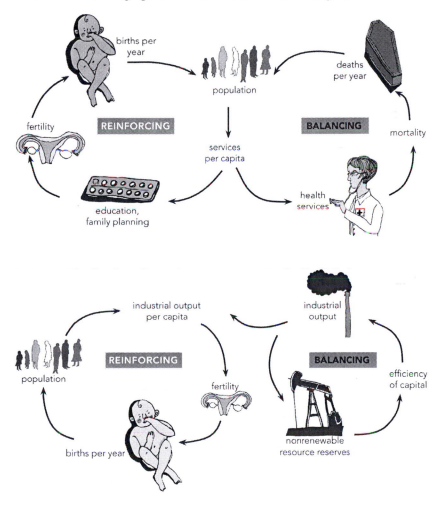

Figure 9.2 *Some important positive (reinforcing) and negative (balancing) feedback systems*

Source: Redrawn by Imogen Shaw based on a figure from *Limits to Growth* (1972)

In a sense, feedback loops represent the opposite logical concept from the simplistic trade-off so beloved of neoclassical economists. Because it uses mathematical methods, neoclassical economics needs to simplify the world before analysing it. These simplifications may reduce the situation under study to just two variables – often illustrated using a two-dimensional x–y graph where one good or service is traded off against another. Thus we could use the same piece of land to produce crops for food or for biodiesel. However, this conceptual paradigm cannot deal with the complex and interconnected systems of ecology, which is one reason why the economic models have failed to predict the serious environmental consequences of exponential economic growth (see Section 2.2). So the simplified trade-off suggests that there is a simple choice between two commodities or situations; however, within a complex ecological system, the commodities and the communities that demand them, and the systems that produce them, are interlinked in complex and sometimes unpredictable ways.

To continue our example, producing food from the land might lead to greater population size, which would mean more people demanding the land for both food and fuel production. Thus the relationship between food production and population is not a simple one of supply and demand reaching an equilibrium; rather, demand may result in more demand. In a similar way, attempts to reduce energy demand by making production processes more energy-efficient may increase the demand for the products themselves, so that ultimately more energy is used (this is called the 'rebound effect': see more in Box 15.1).

The response by economists to the *Limits to Growth* report was muted. This was part of the impetus for the development of ecological economics and its rhetorical assault on economic growth. Herman Daly refers to sustainable growth as 'an impossibility theorem', making reference to neoclassical economist Kenneth Arrow's impossibility theorem of voting behaviour, which was made popular by social-choice theorists. This makes the paper by Arrow and colleagues that was published in the American journal *Science* in 1995 especially striking. The paper attempts to bury the false argument that economic growth actually improves environmental quality because societies need to reach a certain level of prosperity before they are concerned with their environment (the technicalities of this so-called environmental Kuznets curve are discussed in Section 4.3). Rather, its authors argue that economic growth needs to respect the carrying capacity of the planet and the fact that, as the limits of its capacity to absorb pollution are reached, it becomes less resilient and less able to absorb further environmental stress – the most serious example of a positive feedback loop.

Instead of a steadily growing economy, Herman Daly argues the need for a 'steady-state economy':

> By *steady state* is meant a constant stock of physical wealth (capital), and a constant stock of people (population). Naturally, these stocks do not remain constant by themselves. People die, and wealth is physically consumed. That is, worn out, depreciated. Therefore the stocks must be maintained by a rate of inflow (birth, production) equal to the rate of outflow (death, consumption) . . . Our definition of steady state is not complete until we specify the rates of throughput by which the constant stocks are maintained . . . the rate of throughput should be as low as possible.
>
> (Daly, 1971: 29)

Hence, if there is to be growth within the closed system of planet earth then it should be in culture, sociality and leisure, and not in technology or material production. Daly argued that, 'Technology is the rock upon which the growthmen built their church', yet remained unconvinced that technological developments could allow a blithe neglect of ecological realities. A steady-state economy would respect the planet's limits and guarantee that the use of renewable resources did not exceed their ability to be replenished, and that the use of non-renewable resources did not exceed our ability to reuse these. If economic growth requires the use of either sort of resource at a more rapid rate than they can be produced within the ecological system of the planet, then it cannot be compatible with a sustainable economy. Daly saw these as absolute limits, and took issue with the neoclassical theorists and their argument that technological solutions could ensure the production of ever-more material output with lower levels of energy and raw materials.

9.3. Measurement and efficiency

The preceding sections have outlined two diametrically opposed views of the value of economic growth. The source of the conflict can be partially explained by a consideration of what is meant by growth. According to Pearce and Barbier:

> Some of the confusion about the compatibility of economic growth and environmental quality arises from a failure to define concepts and arguments. Thus, some writers regard economic growth as an increase in the materials and energy throughput of the economy. Defined in this

way, the conflict between growth and the environment is not
inevitable . . . But defining economic growth as materials and energy
throughput is not what economists would typically mean by economic
growth. Economic growth is an increase in the level of real GNP over
time. Now the link between growth and environmental degradation is
far less certain because not only can the link between materials/energy
throughput and pollution be broken, but so can the link between
income growth and materials/energy throughput.

(Pearce and Barbier 2000: 29–30)

This brings us neatly to the heart of the disagreement. Can we achieve
economic growth without using materials and energy in such a way that is
impossible with a limited planet? The first and simplest answer is that this
depends on how efficiently we use the materials and energy in economic
production: increasing this efficiency level is known as 'decoupling',
because it decouples economic growth from growth in the use of
materials and especially energy. Following this route towards economic
sustainability is sometimes referred to as 'natural capitalism' (Hawken
et al., 1999). Proponents of this approach suggest that we can improve our
efficiency by a factor of four – producing twice as much using half as
much energy. Lovins gave the example of improving a pipe system by
straightening the pipe and enlarging its diameter, thus reducing the energy
needed to pump fluid through it. Analysis of progress in decoupling
suggests that it has achieved increased energy efficiency in production by
a third (Jackson, 2009: 48), but population increase and greater
consumption have counteracted this improvement.

Hence, while radical critics of the current economic structure would
welcome such efficiency improvements, they would suggest that this can
only postpone and not remove the planetary limit. To achieve a truly
sustainable economy, we need to ask deeper questions about what all the
economic activity achieves, and that process begins by asking what we are
measuring when we measure economic growth. The most popular
measure of economic growth is gross domestic product (**GDP**) – but how
good is GDP as a measure of what we really value, human well-being or
even the well-being of other species and the health of our planet?

GDP measures stocks rather than flows. This has a destructive impact
on the environmental impact of economic activity because a resource that
remains unexploited, e.g. a woodland, has no economic value, whereas if
that woodland is cut down and made into furniture then it acquires
economic value because it can be sold in a market. Hence the depletion
of natural resources has a positive economic impact within such a
measurement system:

> One of the most telling criticisms of conventional economics which environmentalists have been making since the time of the *Limits to Growth* is that in calculating GNP [gross national product] statistics, economists treat the consumption of the Earth's capital as if it were income.
>
> (Dresner, 2002: 75)[11]

Here Dresner is making the same distinction between capital (the woodland) and income (which can be obtained by selling the chairs made from the trees that used to be that woodland). GDP also measures economic activity per se, but makes no attempt to distinguish between activity that is beneficial in terms of well-being and that which is not. An example often cited by critics is that of a major environmental disaster such as an oil spillage following the grounding of an oil tanker. This will add greatly to GDP, since much activity will be required to repair the ship or turn it into scrap, restore the ravaged coastline, treat the injured in hospital, and so on; but the cost of damage to the environment will not be deducted from GDP.

The earliest work undertaken to calculate whether **GNP** is actually a good measure of welfare was conducted by two neoclassical economists, Nordhaus and Tobin (1973). They adjusted the GNP measure in three main respects: distinguishing between expenditure that is only productive and that which relates to consumption or investment in future production; taking into account economic activity that has no value in the market, e.g. housework and leisure activities; and beginning to account for **externalities** as negative in terms of welfare, rather than positive as they are with GNP and GDP. Their conclusion was that GNP correlates fairly well with welfare; this conclusion was disputed by the ecological economists, who developed their own measure, the Index of Sustainable Economic Welfare (ISEW), first devised by Daly and Cobb (Daly and Cobb, 1989). It adds in such factors as domestic labour and the benefits from public services such as street lighting, and deducts value for factors such as habitat destruction and the depletion of non-renewable resources. Figure 9.3 makes clear that human well-being began to diverge from economic activity back in the 1970s, and that our economy has been producing genuine value much less efficiently over time since then. The new economics foundation (nef) has combined this with some measures of well-being to generate its 'Happy Planet Index', which is discussed in Section 9.6. A sustainable economy would need to improve the efficiency with which it uses resources (the ecological footprint is one way of measuring this: see Box 9.1).

Box 9.1

Ecological footprint: useful heuristic or meaningless distraction?

The 'ecological footprint' is a way of thinking about the efficiency with which we use resources to generate economic outputs, and has become very popular. It is a measure of the total amount of land needed to support some given economic activity – so not only to grow food, but also to extract petroleum and turn it into fertilizer to add to the soil, to transport the fertilizer to the farm, to transport the seeds to the farm, to fuel the tractors and other farm machinery, and all the other ancillary demands on the earth that go into the production of that food. When we say that the UK's standard of living would require the produce of three planets to feed all its population, what does this actually mean? Here is how Chambers and colleagues explain the process of calculating the footprint:

Ecological footprint calculations are based on two straightforward assumptions:

1. We are able to estimate with reasonable accuracy the resources we consume and the wastes we generate.

2. That these resource and waste flows can be converted to the equivalent biologically productive area necessary to provide these functions.

Using area equivalence, the ecological footprint aims to express how much of nature's 'interest' we are currently appropriating. If more bioproductive space is required than is available, we can reasonably state that the rate of consumption is not sustainable. Consider a cooked meal of lamb and rice: the lamb requires a certain amount of grazing land, road space for transportation, and energy for processing, transportation and cooking; similarly, the rice requires arable land for production, road space for transportation and energy for processing.

(Chambers et al., 2000: S60)

The authors use the word 'interest' as an analogy with the interest gained from investing money. Critics would question whether the two assumptions are reasonable. Putting a number on our total consumption of resources and production of wastes is a hugely complex task. This must then be translated into an equivalent area of land that is bio-productive in some average sense. For some critics, the process of attempting to calculate ecological footprints indicates the folly of any process of 'costing the earth', while for others it represents a pragmatic response to a policy-making arena in which anything that cannot be accounted for numerically risks being entirely neglected.

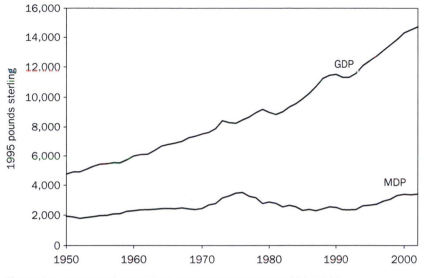

Figure 9.3 *A comparison of GDP and ISEW in the UK, 1950–2002*

Note: ISEW has been relabelled as MDP (measure of domestic progress)
Source: The figure is reproduced with kind permission of the new economics foundation
from Jackson (2002)

9.4. Link between growth and inequality

When GDP figures for countries are given, they are often cited on a per capita basis, which sidelines questions about how economic value is shared. This is of crucial importance, since there is an important link between the concentration of this wealth and the need for our economy to grow – since economic growth can help to ease conflict about allocation. As long as the pie is growing in size, politicians can assuage the demands of those with a smaller allocation by saying that they, too, are becoming wealthier year on year. But once we realize that our survival requires a limit to the size of the pie, the question of how much each citizen receives becomes considerably more urgent. The following quotation makes clear that this connection is recognized at the very top of the economic pyramid:

> Henry Wallich, a former governor of the Federal Reserve and professor of economics at Yale, said: 'Growth is a substitute for equality of income. So long as there is growth there is hope, and that makes large income differentials tolerable.' But this relation holds both ways round. It is not simply that growth is a substitute for equality, it is that greater equality makes growth much less necessary. It is a precondition for a steady-state economy.
>
> (Wilkinson and Pickett, 2009: 221–2)

This quotation comes from a book by researchers concerned with the health effects of inequality, but they have reached the conclusion that an equal society would have to be a society that is in equilibrium with the planet, rather than one that is constantly expanding.

From the other end of the discussion, the ecological economists have also recognized that a steady-state economy will have to be an economy where wealth is equally distributed. In this quotation they make this point by analogy with a badly loaded boat:

> When the watermark hits the Plimsoll line the boat is full, it has reached its safe *carrying capacity*. Of course, if the weight is badly allocated the waterline will touch the Plimsoll mark sooner. But eventually, as the absolute load is increased, the watermark will reach the Plimsoll line even for a boat whose load is optimally weighted.
>
> (Daly and Townsend, 1993: 8)

Chapter 7 presents arguments from those who challenge capitalism on the basis of its lack of justice in allocation, and who see this as related to its environmental destructiveness. However, in discussions around economic growth, we find a growing number of commentators who challenge the distribution pattern of a capitalist economy, although they might not be prepared to question that system as the basic ordering pattern of our economy. The inequality that characterizes capitalism – which anti-capitalists might view as a deliberate mechanism to spur the less well-off to greater efforts as producers and consumers – can be identified by more conventional economists as an aspect of economic life that would no longer be possible in a sustainable economy: 'in a steady-state economy it would not be possible to justify inequality on the grounds that it leads to growth from which all benefit in the end' (Dresner, 2002: 107).

9.5. From quantity to quality

In 1992, Richard Douthwaite published a ground-breaking work, which challenged the economic dogma that growth is the supreme achievement of an economic system. It is a forensic examination of the impact of our expanding economy across a range of areas of life, and it finds that, regardless of questions about whether or not growth can be sustained within planetary systems, growth is simply not increasing human well-being. In fact the reverse: in many ways – from excessively long working hours to rising rates of crime and mental illness – economic growth is creating pressures that are undermining our quality of life while claiming to improve our standard of living:

> For example, higher rates of production at work could affect relationships at home and cause far more unhappiness than could ever be cured by higher wages. The extra production could also increase pollution and cause sickness and misery for thousands of people who could never be compensated adequately from the proceeds of the additional output, even if a way could be found to do so.
>
> (Douthwaite, 1992: 11)

Douthwaite provides compelling evidence that the economic system focused on growth is self-destructive and pernicious in terms of human well-being.

This is not a new discussion: John Ruskin invented a concept of 'illth', which he identified had been produced by the industrial system in equal quantities to the 'wealth' that it was so proud of. That discussion was updated by Herman Daly's concept of 'uneconomic growth', and recently, more mainstream economists have been addressing the same issue, although their concern has more frequently focused on the question of whether the huge levels of economic growth we have seen have actually made anybody happier. Richard Layard, for example, has argued that 'GDP is a hopeless measure of welfare. For since the War that measure has shot up by leaps and bounds, while the happiness of the population has stagnated' (Layard, 2003: 3). Layard's conclusion is that, while a certain minimum income standard for individuals and societies is a necessary condition for happiness, beyond a certain level of growth the negative impacts of excessive economic activity – whether in terms of pollution or the breakdown of relationships that can result from working too hard – begin to outweigh the positive. In other words, even leaving aside the question of the impact of economic growth on the environment, we need to find a way of balancing economic growth with human well-being.

Layard was appointed a 'Happiness Tsar' by the UK government, and another report to that country's government suggests this may be the beginning of a policy trend: the UK Sustainable Development Commission recently produced a report called *Prosperity Without Growth* (Jackson, 2009), which makes a similar case. The Commission argues that we need to develop policy to support 'flourishing within limits': learning to value what we have rather than always seeking more. They also suggest a renewed emphasis on social, rather than material, well-being with more emphasis on the value that we derive from social interactions. They use the phrase 'alternative hedonism' to suggest a way of life that focuses on sensual pleasure derived from low-energy and low-consumption lifestyles, such as the pleasure that comes from truly appreciating our local environment and community (Soper and Thomas, 2006). There is a whole movement now dedicated to developing this sort of more humane economy, which is called the Degrowth Movement. Its slogan – 'Moins de biens; plus de liens' or 'More fun, less stuff' – sums up the shift in emphasis from material goods to human relationships that is designed to counter the growth dynamic that currently drives our economy (Fournier, 2008).

9.6. Case study: Growth and happiness in global perspective – the nef's Global Happiness Index

If economic growth is not intended to increase human well-being or happiness then what good is it? That is the question that has spurred London-based research institution, the nef, to produce its Happy Planet Index. It compares the happiness levels of people in different countries with the environmental impact they caused in achieving that happiness level – and generates some interesting and surprising results.

The formula used to calculate the Index is as follows:

$$HPI = \frac{\text{Life satisfaction} \times \text{Life expectancy}}{\text{Ecological Footprint} + \alpha} \times \beta$$

This formula basically represents a ratio of the quality of life to the environmental cost with which it was bought (two Greek letters are included to make the data easier to compare). While life expectancy is a relatively easy number to ascertain, the figures for life satisfaction are obtained from surveys, where people are asked to assess how happy they are – there is thus a considerable problem with comparisons across countries. More details on how the numbers were calculated are available on the nef website (http://www.neweconomics.org). The first Index was calculated for 177 countries, of which the Pacific island group of Vanuatu scored highest (ironically, Vanuatu is threatened with inundation resulting from climate-change-related sea-level rises) and Zimbabwe scored lowest. The results for 2009 indicate particularly good scores for the countries of Latin America. (Table 9.1 reports the 2009 Happy Planet Index values for a range of countries.) Comparisons can also be very revealing, for example between the USA and Germany, which have similar levels of life satisfaction and life-expectancy. However, the ecological footprint for Germany is around half the size of that of the USA, which indicates that

Germany is far more efficient at producing satisfied citizens. As the nef concludes:

> The message is that when we measure the efficiency with which countries enable the fundamental inputs of natural resources to be turned into the ultimate ends of long and happy lives, all can do better. This conclusion is less surprising in the light of our argument that governments have been concentrating on the wrong indicators for too long. If you have the wrong map, you are unlikely to reach your destination.
>
> (http://www.happyplanetindex.org/learn/what-hpi-tells-us.html, accessed 13 September 2010)

Table 9.1 A selection of countries ranked by HPI values, 2009

Rank	Country	Life expect.	Life satisfac.	Footprint	HPI
1	Costa Rica	78.5	8.5	2.3	76.1
2	Dominican Republic	71.5	7.6	1.5	71.8
3	Jamaica	72.2	6.7	1.1	70.1
9	Brazil	71.7	7.6	2.4	61.0
12	Egypt	70.7	6.7	1.7	60.3
20	China	72.5	6.7	2.1	57.1
35	India	63.7	5.5	0.9	53.0
43	Netherlands	79.2	7.7	4.4	50.6
51	Germany	79.1	7.2	4.2	48.1
53	Sweden	80.5	7.9	5.1	48.0
71	France	80.2	7.1	4.9	43.9
74	UK	79.0	7.4	5.3	43.3
75	Japan	82.3	6.8	4.9	43.3
89	Canada	80.3	8.0	7.1	39.4
102	Australia	80.9	7.9	7.8	36.6
114	USA	77.9	7.9	9.4	30.7
115	Nigeria	46.5	4.8	1.3	30.3
142	Tanzania	51.0	2.4	1.1	17.8

Source: The Happy Planet Index 2.0: Why good lives don't have to cost the Earth (London: new economics foundation)

Data sources: Average life expectancy at birth was taken from the 2007/8 Human Development Report of the UN; life satisfaction data from Gallup's World Poll and World Values Survey; ecological footprint data from the World Wildlife Fund's Living Planet Report 2008

Summary questions

- Why would a neoclassical economist be unconcerned by the suggestion that we may have no further stocks of petroleum within 50 years?
- Can you think of increases in three types of economic growth that would not offend Herman Daly?
- What flaws can you see in the calculation of the Happy Planet Index?

Discussion questions

- If climate change leads to more freak weather events, will this add to, or subtract from, economic growth?
- Why would people be more concerned about equality in a steady-state economy?
- Does it matter whether the ecological footprint is made on land in the Sahara Desert or the plains of the Middle West USA?

Further reading

Douthwaite, R. (1999), 'The Need to End Economic Growth', in Cato, M. S. and Kennett, M. (eds), *Green Economics: Beyond Supply and Demand to Meeting People's Needs* (Aberystwyth: Green Audit): a brief introduction to why green economists reject the need for economic growth, and which types they might countenance.

Nordhaus, W. and Tobin, J. (1973), 'Is Growth Obsolete?', in Moss, M. (ed.), *The Measurement of Economic and Social Performance: Studies in Income and Wealth*, vol. 38 (New York: Columbia University Press), pp. 509–64: offers a view of the same debate from a more mainstream position.

Pearce, D. W. and Barbier, E. B. (2000), *Blueprint for a Sustainable Economy*. (London: Earthscan), especially ch. 2: two environmental economists' take on the same issue.

Weizsäcker et al (1997), *Factor Four: Doubling Wealth – Halving Resource Use: The new Report to the Club of Rome* (London: Earthscan): a useful introduction to the possibilities for decoupling economic activity from energy use.

10 All that the earth provides: the economics of resources

In the previous chapter we considered, from a theoretical perspective, whether the process of economic growth faces any limits. This chapter and the following one will examine two specific forms of limitation. In this chapter, we will address the issues of inputs to the production process, and how, on a limited planet, we should approach the issue of limited natural resources. The following chapter will address the other end of the production process – the issue of how we deal with the wastes it produces. Section 10.1 explains how scarcity is a central guiding concept in orthodox economic theory, linking the availability of resources to their market prices. Section 10.2 draws insights from environmental economics, extending the theory of the relationship between prices and scarcity by considering whether creating markets can help to preserve one particular scarce resource: the rainforests. Section 10.3 provides a critical view of this commodification of natural resources. Section 10.4 provides another take on scarcity, by economists who consider that the earth is abundant and that it is our market-based perceptions that are limited. Finally, Section 10.5 offers a case-study of different approaches to protecting the world's fisheries.

10.1. A science of scarcity

Scarcity is *the* central proposition of mainstream economics, which is defined (following Lionel Robbins) as the study of the allocation of scarce goods among competing ends. The *Oxford Dictionary of Economics* uses a slightly different definition: 'The study of how scarce resources are, or should be, allocated'. The distinction between a scarcity of resources and a scarcity of goods is quite an important one, since whether or not goods are scarce depends on people's subjective perceptions of their needs and desires. The scarcity of resources, by contrast, is fixed by the level of those resources available on this planet – at least until (or unless) we can find a way to access resources elsewhere in the universe. It is important to understand the difference between scarcity and limitation: a scarce resource is one that is limited *relative to* the amount of it that is sought; this is distinct from a resource that is naturally limited. As a broad generalization, we can say that neoclassical and environmental economists focus on scarcity, whereas ecological and green economists are more concerned with resource limits.

From both perspectives, however, we see that the role of economics is to find mechanisms, and perhaps moral values, to decide how the limited resources or goods are shared out. Economic theory is based on an assumption of scarcity, and then suggests that markets are the most efficient method for determining this allocation. Markets function to distribute scarce resources according to a supply-and-demand curve like that illustrated in Figure 10.1. The axes represent price (on the y-axis) and quantity (on the x-axis); this graph relates to the price of natural gas in the world market, but it could apply to any individual resource or good. The fact that the demand curve slopes downwards from left to right indicates that, as the price rises, less of the good will be demanded. Conversely, the supply curve slopes upwards, indicating that, as the price of a good rises, more of it will be supplied. The assumption is that the good is scarce: if it were freely available, it would have no price.

The market mechanism maximizes utility since if the price is too high (point P_0 in the figure), then more gas would be supplied than is demanded (Qs – Qd). In order to sell their production firms will lower their prices (to Pe, the equilibrium price). From the customer perspective, at price P_0 they would only demand quantity Qd, leaving surplus production which would not be demanded until the price reaches Pe. So the market operates to negotiate the price between the producer and the consumer, tending towards an equilibrium point Qe.

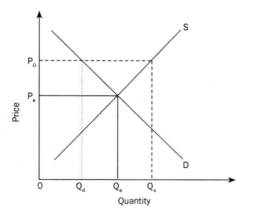

Figure 10.1 *How the market sets the price of a scarce resource*

Source: Based on Hussen (2000), Figure 2.5; redrawn by Imogen Shaw

As well as the absolute availability of a natural resource, two other factors influence the price: the availability of substitutes and the level of existing technology. Before we had natural gas, coal was used to make town-gas, but natural gas was cheaper, and hence 'gas works' fell into disuse. However, if competition for sources of natural gas were to lead to significant price rises on global markets, countries with large coal reserves could again use them to extract gas. Or perhaps other technologies will be developed which will allow us to create alternatives to natural gas that we cannot even imagine yet. Alternatives to fossil fuels for generating electricity are already available – wind power and nuclear power, for example – and these also influence the price of fossil fuels. The price of extraction is clearly related to availability: as sources of the resource that are easy to extract are exhausted, the price rises, and other sources then become economically viable. An example of this process is the move from the exploitation of petroleum reserves that can be drilled and pumped in the Middle East or the USA, to reserves which are heavier, harder to extract, more energy-intensive to extract, and more expensive to refine – such as those found in the Athabasca tar sands of North-Western Canada (see Section 11.5). Thus price results from the complex relationship between the quantity of a limited resource that is physically available and our ability to access that resource – what Hanley et al. (2001: 321) refer to as 'a race between depletion and technological change to keep costs down'.

So far this has been a fairly theoretical discussion. We are now going to enliven it by considering an old debate that has a contemporary feel: Jevons's concern that coal, on which the Industrial Revolution and British global domination was based, might be a finite resource. Jevons wrote his report in 1865 – the year that marks the beginning of the expansion of the South Wales coalfield, which suggests that he was unduly pessimistic – but it is still a ground-breaking work for taking seriously the issue of resource depletion and its impact on economic activity. One example of the sophistication of Jevons's thinking about resources is his creation of the 'Jevons Paradox' – that as our use of energy becomes more efficient, the cost of energy falls, and therefore we are inclined to use more. This might apply to our current use of low-energy light-bulbs – because we know they are not costing us so much, we may leave lights on for more of the time (see more on these 'rebound effects' in Box 15.1). Jevons's theories about resource depletion and its impact on the economy have also recently been adopted by those concerned about 'peak oil' and the inevitable decline in oil supplies. Thus a theory that was developed 150 years ago in terms of one fossil fuel may be relevant in our own times to another, when we face a decline in the easy availability of petroleum, which is the resource that underpins the complex global market we rely on for the majority of our basic resources.

Jevons has been adopted by the ecological economists, because of his focus on the limits to growth of the macroeconomy that the exhaustion of coal reserves might enforce. Herman Daly's extension of this thinking to include his own concept of 'uneconomic growth' is illustrated in Figure 10.2. The rightward-sloping curve above the x-axis illustrates the benefits of growth, while the broken curve represents the cost of that growth – the impact that the economic activity has on the ecosystem. Daly's concept of

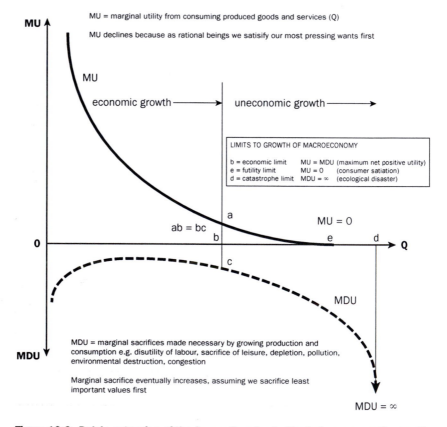

JEVONIAN VIEW OF LIMITS TO GROWTH OF MACROECONOMY

MU = marginal utility from consuming produced goods and services (Q)

MU declines because as rational beings we satisify our most pressing wants first

economic growth ⟶ uneconomic growth ⟶

LIMITS TO GROWTH OF MACROECONOMY

b = economic limit   MU = MDU (maximum net positive utility)
e = futility limit   MU = 0   (consumer satiation)
d = catastrophe limit   MDU = ∞   (ecological disaster)

ab = bc

MU = 0

MDU = marginal sacrifices made necessary by growing production and consumption e.g. disutility of labour, sacrifice of leisure, depletion, pollution, environmental destruction, congestion

Marginal sacrifice eventually increases, assuming we sacrifice least important values first

MDU = ∞

Figure 10.2 *Daly's extension of the Jevons Paradox to illustrate uneconomic growth*

Source: First Annual FEASTA Lecture, 26 April 1999 (Dublin: FEASTA), available online here: http://www.feasta.org/documents/feastareview/daly.htm.

uneconomic growth arises when the costs of economic activity outweigh the benefit – any point beyond b on the x-axis, where ab = bc, or the costs and benefits of the economic activity are equivalent. Once an economy grows beyond that point, the extra economic growth is of less value than the damage the activity has on the ecosystem. By the time you get to point e, you have reached what Daly calls the 'futility limit' – you have so much stuff that you no longer have time to enjoy it! Point d represents 'the catastrophe limit', where an impressive new invention turns out to have terminally destructive impacts on the ecosystem – perhaps a genetically modified crop including a terminator gene that spreads to all plants and thus destroys the process of photosynthesis and all life on earth.

10.2. Using prices to protect natural resources

'Can orthodox economics save the forests?', asked David Pearce (1998: 193). He was writing in the context of his urgent concern with the rapid loss of tropical rainforest that had been building since the 1980s, when they were recognized as important sources of resources and valuable plant and animal species, as well as massive 'sinks' for carbon dioxide. The loss of forest has accelerated since then (as illustrated in Figure 10.3); the reason for this loss is a source of debate between economists.

Pearce explains the economic failure that has led to the loss of rainforest, as detailed in Box 10.1. The assumption behind this explanation is that the market is an ideal form of allocation of goods and services, even rather abstract ones such as biodiversity. If the market has failed to protect

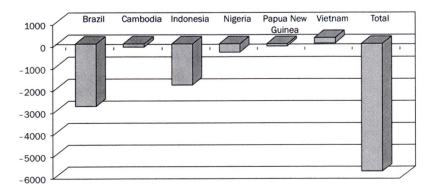

Figure 10.3 *Loss of tropical forest, 1990–2005 ('000ha.)*

Source: Author's graphic; data from the Mongabay website (www.mongabay.com), derived from the FAO, Forest Resources Assessment 2005

the rainforest and its species adequately, that is because the market is not functioning efficiently, rather than because the market does not function well in such fields, which might better be the preserve of political decision-making. Pearce makes a strong case against such a view. He identifies the competition for space between natural species and human beings as the key problem. In order to solve the problem, he suggests that more understanding is needed and that this should be achieved by using complex mathematical techniques (known as 'regression analyses') to determine which factors (or 'explanatory variables') explain the 'dependent variable', which is rainforest loss.

Box 10.1

Sources of economic failure to prevent the destruction of tropical rainforest

Cause	Explanation
Local market failure	Property rights are poorly defined: because nobody owns the forest, no one has the incentive to protect it.
'Missing markets' for rainforest products	Goods that might be traded and therefore protected are not actually being bought and sold, and biodiversity is therefore being squandered.
'Missing markets' for non-use values of maintaining biodiversity	Although valued by other countries, it is not commodified so that they cannot indicate their preference for it with money.
'Missing market' for the carbon storage potential of the forests	Since this property right is not defined, it cannot attract foreign investment and therefore has no value to the national governments concerned.
High transaction costs	The expense of taking the conservation benefits of the forest to market is prohibitive, which leads to their under-exploitation.
Public goods problems	The conservation benefits are shared by everybody, so nobody has an incentive to pay for them.
Intervention failure	The Brazilian government, for example, had been paying subsidies to farmers to convert forest to livestock grazing at least until the end of the 1980s.

| Poor policy choices creating the wrong incentives | The government had also imposed inadequate taxes on logging companies and encouraged inefficient wood-processing companies who wasted much of the wood resource they harvested. |
| Global market failure | Property rights to the global atmosphere are poorly defined, and hence nobody has an incentive to pay for the CO_2 absorption service offered by the forests. |

public good

Source: Pearce (1998)

The results of such regression analyses indicate that population growth and population density are both clearly linked to the loss of forests. Rising rates of income can also help to explain an increase in deforestation. For two other possible explanatory variables – agricultural productivity and indebtedness – the results were inconclusive. While one might expect higher levels of agricultural productivity to reduce pressure on the forests, half the results point in the opposite direction. No significant link can be found between a country's level of indebtedness and the likelihood that it will lose its tropical forest.

Pearce concludes that orthodox economics can save the rainforests so long as it manages to demonstrate the huge amount of economic value that lies in those forests, and if it can create mechanisms for enabling trade between those who wish to save the rainforests and those who are responsible for their management. He suggests the need for 'an imaginative use of a wide range of instruments – from **debt-for-nature swaps** to tradable development rights to joint implementation and private green image investments' (see also Table 4.1). His conclusion is that orthodox economics should be given a chance because 'the unorthodox is without appeal or practicality. And the non-economic approaches have failed' (Pearce, 1998: 205–6). All these suggested policies rely, of course, on being able to generate monetary values for various aspects of the rainforest that can be considered worth protecting. Some attempts to put such a value on the natural world through a technique known as 'willingness to pay' (WTP – see Section 4.2 for a fuller explanation of these valuation methods) are reported in Table 10.1.

Table 10.1 Debt-for-nature swaps: implicit global willingness to pay via international transfers for the protection of tropical forest in a range of countries

Country	Date	Payment	Area (m.ha)	WTP/ha (US$1990)
Bolivia	Aug. 1987	112,000	12.0	0.01
Ecuador	Dec. 1987	354,000	22.0	0.06
Costa Rica	Feb. 1988	918,000	1.15	0.80
Philippines	Jan. 1989	200,000	9.86	0.06
Madagascar	July 1989	950,000	0.47	2.95
Nigeria	1989	1,060,000	1.84	0.58

10.3. The planet sold to service our desires

There is no argument amongst the economists whose contributions are included in this book that the planet is valuable – it is the question of what that value means and how it can be measured that is the source of disagreement. For environmental and ecological economists, it makes sense to attempt to introduce a monetary valuation of the planet, and to establish property rights to protect this value. Without the earth, we would not survive as a species. We breathe air, eat food produced from soil, use natural materials for buildings – the earth provides the feedstocks for every productive process, and without these resources we could not survive. Just how valuable these resources are has been measured:

> People still fail to understand just how big a risk this is. Back in the 1990s when Bob Costanza and his colleagues at MIT did their calculations on the total monetary value of the 17 principal 'eco-system services' on which we depend (things like building fertility in the soil, flood control, climate regulation and so on), the figure they came up with was $33 trillion – pretty much the same economic value as one year's worth of GDP.
>
> (Porritt, 2009: 20)

This concept of **ecosystem services** is a fairly recent one in economic debates and has been received somewhat controversially:

> What are 'Ecosystem Services'? At first hearing, they sound like a firm of consultants who help you repair your ailing ecosystem. In fact it's the other way round, the service is provided by people with ecosystems to people who no longer have one, and who need one. For example if your forest, or your peat bog is absorbing carbon, it is providing a service to other people who are producing excessive CO_2

and need something, somewhere to absorb it. Other ecosystem services include climate regulation, maintenance of biodiversity, water conservation and supply, and the preservation of aesthetic, cultural and spiritual values. The emerging view is that the people receiving these ecosystem services should start to pay for them.

(Sullivan, 2008a: 21)

Ecosystem services formed a major focus of the recent study known as the Millennium Ecosystem Assessment (MA), which reported that '60 to 70 per cent of our world's ecosystem services are deteriorating, with dramatic consequences for those who are most dependent on their steady provision, such as subsistence farmers'. There is an explicit admission that the concept of ecosystem services has been designed to increase the attractiveness of talk of environmental protection to the corporate sector:

The attractiveness of the 'ecosystem services' concept is also largely due to its capacity to provide a unifying language between the economic, business and environmental communities; as beneficiaries of valuable services are identified, previously uninvolved actors are recognizing that they have a stake in conserving the environment.

(UNEP, n.d.: 2)

However, it may be an important cost if the planet's intrinsic spiritual value is lost in the process of 'costing the earth'.

There are also difficulties in terms of establishing ownership rights over the areas of the world where ecosystems remain intact, largely due to low levels of industrialization. Since the people living in these areas have less economic clout, they may not be well-placed to protect their rights over their land and their lifestyle – which, paradoxically, is precisely what has preserved the ecosystem. As environmental pressures increase, subsistence farmers in the world's poorer nations are threatened with displacement and loss of livelihood as their land is traded to provide **carbon sinks** and other ecosystem services for the peoples of the richer world.

It may seem strange to think of the planet in terms of a range of services, and it seems to make little sense to attempt to measure them because their value is clearly infinite. Without the planet, we cannot survive, so all the money we could ever create would not be a large enough price to pay. From a green economic perspective, this whole discussion would appear to suffer from the familiar problem of a category error, i.e. considering the ecosystem as part of the economy rather than the economy as an entirely dependent part of the ecosystem.

10.4. Scarcity or abundance

Green economists have a tendency to go right back to the basics of the discussion around the exhaustion of mineral or natural resources, and to criticize the orthodox view of scarcity on the basis that it leads to a distortion of our relationship with nature and that focusing on scarcity can actually generate destructive behaviours. For example, if you believe that a resource is scarce, your tendency will be to accumulate it, or the power to access it in future. Hence, you may well over-exploit the resource simply as a result of your fear that it may become exhausted. An early book that was very influential on the development of green economics challenges the scarcity basis of economics in its title, *Wealth Beyond Measure* (Ekins et al., 2000).

In this thinking, green economists follow a tradition established by Karl Polanyi, an economist who has not received as much attention in this book as he might have done, largely because he does not fit into any particular school of thought. Polanyi's interest was in pre-modern, pre-capitalist societies and how they dealt with the problem of allocation. He viewed capitalism as a recent – and regressive – economic development in the long history of human society. Polanyi (1945) considered the focus on scarcity as a symptom of a capitalist economy; pre-capitalist economies, based on systems of reciprocity rather than competition, had little need for such a concept.

Marshall Sahlins (1972) takes this thinking a stage further, in contrasting the levels of abundance of modern, Western societies with those of hunter-gatherers, which he characterized as 'the original affluent society'. He suggested that the focus on material wealth is a symptom of an unhealthy society that lacks an ability to judge what is really valuable in life, a system of thought he refers to as 'the Zen road to affluence'. This relates back to a point made at the beginning of this chapter, about the difference between scarcity and limits. While focusing on scarcity may encourage us to consume ever more, a focus on abundance may help us to experience satisfaction with what we have, and therefore consume less. Thus scarcity may be an attitude of mind – driven by the nature of our culture and its ethical values – rather than an absolute concept. This approach is encouraged by green economists, who believe that it may help to guide our way to an economy that is able to live within natural limits.

Other radical green and anti-capitalist economists go further, suggesting that scarcity and poverty are constructs, deliberately constituted by the capitalist economic system, which depends on the motor of accumulation to motivate people to sell their labour in the workplace. The early critics of capitalism thus saw:

> The glaring paradox between the unprecedented productive potential of industrial capitalism in creating the most materially affluent societies the world has ever seen alongside the persistence (and indeed deepening) of poverty within such materially affluent societies . . . For many critics of the new capitalist social order, it was evident that it could not eradicate poverty, since the economic system required the poverty of the mass of its population in order for them to engage in wage labour.
>
> (Barry, 2007: 220)

This argument relates to the link between economic growth and inequality that is discussed in the previous chapter. The struggle for value between workers and those who extract some of the value of their work in profit can be assuaged, to some extent, by the use of more resources and more efficient technologies to increase the overall quantity of material outputs that are available to be shared. Of course, this is also an explanation for the over-use of those resources, as well as a stimulus to technological innovation.

10.5. Case study: Fisheries policy

So much for the theory. How does it apply to one threatened resource, the global fisheries? Neoclassical economics approaches this problem by devising a theoretical model for the supply and demand of fish, such as that illustrated in Figure 10.4. The graph illustrates the return that a fisherman gains compared with the effort he puts into fishing. The semicircle represents the amount of effort put into the fishery as a whole, which increases as more fishermen use the fishery because returns are good, but then declines because reducing stocks mean there are no longer profits to be made, so fishing effort declines. The cost of that effort itself increases steadily, as represented by the 45-degree line from the origin; the MEY (maximum economic yield) line is parallel to this line. E1 is the point of maximum profit, where the difference between costs and revenue is greatest. Point E2, where total revenue is equal to total cost, represents the beginning of the decline of the industry. After that point, there is nothing to be gained from further fishing. The ideal point from the social perspective is E, but if there is a situation of competition between fishermen then the market will reach equilibrium at point E2, which is socially inferior, with the same number of fish being caught at greater cost.

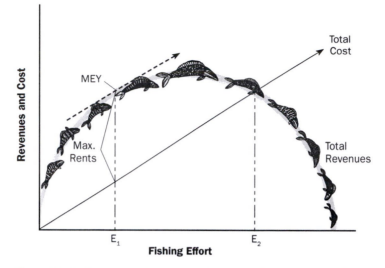

Figure 10.4 *The economic situation facing a fisherman: reward in return for effort*

Source: Drawn by Imogen Shaw

At least for the case of just two fishermen, neoclassical economics has a system for dealing with the sort of negotiation between fishermen to achieve an optimal solution that is preferable to the competitive market solution, a solution derived from game theory (this is based on the account in Hanley et al. (2001: 154–6)). The basic proposition of game theory is derived from the so-called 'prisoner's dilemma', in which two prisoners have collaborated in a crime and are arrested and questioned in separate cells. The police do not have enough evidence to convict either and offer both, separately, the same deal. If one prisoner offers evidence against his partner in crime then he will be freed but his partner will receive a tough sentence. Each faces the same possible outcomes: if one prisoner betrays his partner and his partner stays loyal he will be let go; however, if both cooperate with the police then they will be punished, but less severely; if neither betrays the other then they cannot be convicted of the main charge for lack of proof. However, both prisoners with recieve a moderate sentence for a more minor crime they have committed, which the police have evidence for. Each prisoner finds it rational to strike a deal as this always improves their outcome regardless of their partner's actions. However, each must trust that the other will not offer evidence against him, which would result in the worst outcome. The point of the game is that the best outcome arises from trust and cooperation, but there is also a temptation to fear that your accomplice will confess and therefore to arrive at a worse outcome. In terms of the problem of a shared fishery, it is also clear that cooperation between fishermen would produce an outcome that would be better for all – and especially the fish! – than a competitive, market response.

In the EU, fisheries are governed by the Common Fisheries Policy, which has been an example of disastrous regulatory failure. During the 1990s, it became obvious that fish stocks in the waters around Europe were declining massively. EU policy on fisheries was dominated by the perspective of neoclassical economics, which suggested that this was a problem of 'the tragedy of the commons' (see the account in Section 14.1) and that the solution must be to introduce clear property rights into the situation. The fisheries policy that was developed was based on three principles: the scientific determination of a total allowable catch (TAC) based on an estimate of remaining fish stocks; national quotas allocating this limited catch between EU members; and trading in these quotas between individual fishermen. This is a classic example of a market solution to an environmental problem; unfortunately it has not been effective.

The market turns out to be a far-from-competitive one. Large, corporate fishing fleets are able to dominate the fisheries. They also use methods and equipment that allow them to hoover up the content of the sea rather than catching particular fish of a certain size. Because the quotas are specific about species and size, this leads to massive amounts of dumping of 'by-catch' – fish that have been caught but cannot be landed legally. So the declining levels of fish landed have only succeeded in destroying the livelihoods of fishing communities, and not allowing regeneration of stocks because they do not reflect more live fish being left in the sea, only dead ones that are discarded:

> With relative free-for-all unleashed, law-abiding fishers found themselves criminalized as they struggled to compete with continental fleets and out-of-touch regulations. Most notorious of all is the enforced dumping of unwanted accidental by-catches of species that exceed quotas. As one skipper said 'I've just returned from yet another fishing trip where we were forced to dump 200 boxes of coley and 100 boxes of haddock—values up to £18,000. I'm absolutely disgusted with this total waste of resources.'
>
> (McIntosh, 1998: 14)

Fishing consultant David Thomson (2006) estimates that up to 600,000 tonnes of fish from all boats operating in British waters are discarded in this way every year. For comparison, British fishing boats landed 614,000 tonnes of fish in 2006. In other words, about the same amount of fish is dumped as is brought ashore. The EU estimates that 40–60 per cent of the North Sea catch is rejected due to not meeting a quota. Some of Thomson's evidence to a UK House of Lords enquiry into the impact of the EU Fisheries Policy on local communities is reproduced in Box 10.2. It was influential in persuading the Scottish government to change its policy and introduce a Cod Recovery Plan, aiming to reduce the level of discards. This has been influential on the wider EU policy, which is now more focused on the needs and responsibilities of fishing communities. It takes an economic view more like that of the political scientist Elinor Ostrom (1990) who observes that those whose livelihoods depend on common resources can best develop institutions to protect them (see Figure 10.5).

Box. 10.2

The view from the fishing communities

A review of the past 30 years, or just the period in question since 2002, reveals that the impact of the EU Common Fisheries Policy management on UK fisheries has been almost entirely negative. Whether we look at annual stock assessments, continuing fleet decline or the stagnation of the economies of coastal fishing towns and villages, there is almost no sign of recovery.

TACs are based on the scientist's best estimate of stock size, plus the judgement of fishery administrators and also the lobbying of national governments for a greater share for their respective fleets. The end results rarely satisfy any party. The estimates are based on two-year-old data, and in some cases on bogus statistics. The colossal amount of discards – up to 600,000 tonnes per year – was never calculated accurately.

Source: http://www.alastairmcintosh.com/general/resources/
2008-David-Thomson-Fishing-Industry-Community.pdf
(accessed 13 September 2010)

	LARGE SCALE 	SMALL SCALE 
Number of fishermen employed	AROUND 500,000	OVER 12,000,000
Annual catch of marine fish for human consumption	AROUND 29 MILLION TONNES	AROUND 24 MILLION TONNES
Capital cost of each job on fishing vessels	$$$$$$$$$$$$$$$$$$$$$$$$$$$$$$$ $$$$$$$$$$$$$$$$$$$$$$$$$$$$$$$ $$$$$$$$$$$$$$$$$$$$$$$$$$$$$$$ $$$$$$$$$$$$$$$$$$$$$$$$$$$$$$$ $ 30,000 - $ 300,000	$ $ 250 – 2,500
Annual catch of marine fish for animal feed and industrial reduction to meal and oil.	AROUND 22 MILLION TONNES	ALMOST NONE
Annual fuel oil consumption	14 - 19 MILLION TONNES	1 – 2.5 MILLION TONNES
Fish caught per tonne of fuel consumed	= 2 – 5 TONNES	=
Fishermen employed for each $ 1 million invested in fishing vessels	5 – 30	500 – 4,000
Fish destroyed at sea each year as by-catch in shrimp & trawl fisheries	6 – 16 MILLION TONNES	NONE

ICLARM: David Thomson's illustration above created widespread awareness of the efficiency of small-scale fisheries; however, some donor agencies still feel obliged to "upgrade" them into inefficient large-scale fisheries! The table above has been brought up-to-date by courtesy of Dr. Armin Lindquist, Assistant Director-General (Fisheries Department), using latest (1986) FAO fisheries statistics and economics data at 1988 prices. (table re-formatted by Kim Ang, FAO project I.T. mapping and data specialist, 2007)

Figure 10.5 *The impact of small-scale vs. large-scale fisheries in terms of job creation, environmental impact and fuel use*

Source: Based on FAO data and reproduced from Thomson (2006)

Summary questions

- Why was Jevons wrong about the exhaustion of coal supplies?
- How do you understand the concept of 'ecosystem services'?
- How useful is a measure of the global value of the ecosystem? What type of money should we use to price it in?

Discussion questions

- Where do you think your perception of scarcity comes from?
- Is the failure of EU fisheries policy the result of too much government or too much market?
- Are you more likely to protect the planet if you have a perspective of scarcity, or a perspective of abundance?

Further reading

Ekins, P., Hillman, M. and Hutchison, R. (2000), *Wealth Beyond Measure: An Atlas of New Economics* (London: Gaia Books): an excellent early introduction to a green approach to the issue of scarcity vs. abundance.

Hussen, A. M. (2000), *Principles of Environmental Economics: Economics, Ecology and Public Policy* (London: Earthscan), chapters 2 and 3: covers the issues of resource scarcity from an environmental economics perspective.

Sullivan, S. (2008b), 'Global Enclosures: An Ecosystem at Your Service', *The Land*, Winter, 21–3: a short, sceptical account of recent attempts to price the earth and its services.

 Pollution

In the previous chapter, we looked at the impact of planetary limits on the inputs to production processes. In this chapter, we will look at the same issue from the other end by addressing the outputs from production processes that have to be assimilated by the environment. From a conventional economics perspective, pollution is an inevitable part of economic life. As long as we are engaged in transforming material inputs (raw materials and energy) into economic goods, we cannot avoid creating residuals. One of the most important concepts underlying economics is the trade-off, which is implicit in Milton Friedman's famous statement that 'there is no such thing as a free lunch': you cannot receive a benefit without sacrificing something in return. Section 11.1 discusses neoclassical economics' view of pollution as part of a trade-off – if we wish to have the goods then we will have to endure the pollution that their production generates. Section 11.2 then moves on to explore a regulatory approach to tackling pollution, which begins from the assumption that it is a social and political, rather than a market, problem.

The residuals of production processes can be considered to be 'pollution': an inevitable and acceptable by-product of production processes, which has to go somewhere – buried in the landscape, thrown into the sea, or released as gas into the air. The natural environment can – given enough time, and if it is not overwhelmed – break down these wastes, and hence an ecological approach to the economy would work more closely with nature, as outlined in the approach of industrial ecologists, whose contribution is covered in Section 11.3. Section 11.4 asks fundamental questions about what pollution is, and what it is about our economy that generates it in the first place. Finally, in Section 11.5, we consider one specific example of pollution: the pollution to local waterways caused by the exploitation of the Athabasca tar sands in Canada.

11.1. Neoclassical economics: internalizing the externality

According to the neoclassical view, while pollution is a by-product of industrial production, when it reaches the level at which it causes unacceptable social consequences it is an example of market failure, that is to say if the market were functioning optimally the problem would not have arisen. So what is the failure? While markets may be an efficient way of allocating goods between individual consumers, they can face problems with negative impacts at the level of society as a whole. Economists define these negative social consequences of economic activity as **externalities**. For a conventional economist, an externality exists 'when the market price or cost of production excludes its social impact, cost, or benefit'. Because it is possible for the producer to generate pollution that it does not pay for itself, the price it charges to consumers does not accurately reflect the price of production. The cost of the pollution is external to the price- and cost-setting within the firm or industry. In standard economics, the price and quantity of production are set by finding the equilibrium point between supply and demand curves. The supply curve responds to the costs faced by suppliers. If they are able to externalize some of these costs (those associated with the pollution) then the efficient equilibrium point is not reached – the supplier is over-producing.

This is illustrated by a series of figures built on the classic supply-and-demand analysis. It should first be noted this assumes that some pollution is inevitable – the objective should be to produce an efficient outcome, i.e. one that maximizes the utility of society as whole. The graphs are used to determine what the optimal level of production of the good in question might be. This is determined by balancing the utility that results from the production of the good with the disutility caused by the pollution that results as a by-product of the production process. When describing pollution, the standard supply and demand curves are replaced with a 'marginal pollution cost' (MPC: the supply curve) and a 'marginal social benefit' (MSB: the demand curve). Economists are interested in what happens 'at the **margin**', i.e. at the point of change from something being to your advantage to it being to your disadvantage. In the case of price-setting by firms, this would relate to marginal cost, i.e. the amount the firm would have to pay to produce one more item of production, for example a car. This is clearly different depending on whether the firm has already made 10, 100 or 10,000 cars. Hence marginal cost is related to volume because many of the costs of production – renting the factory, paying staff, acquiring expertise, and so on – are fixed regardless of how many cars you produce. When thinking about pollution, the focus of attention is the marginal social benefit, i.e. the advantage to society of one less kilogram of pollution being produced and the marginal pollution cost, i.e. the cost to the firm of treating its effluent so that this last unit of pollutant is not emitted.

The first graph (Figure 11.1) illustrates the 'consumer surplus' when we only consider the private costs of the production process in terms of a classic demand-and-supply graph. Supply is represented by the line that is rising at a 45-degree angle from left to right; demand is represented by the line that declines from left to right. The y-axis indicates the price and the x-axis the quantity bought and sold. The basic graph thus illustrates the relationship between price and quantity demanded, so that the higher the price, the fewer consumers that wish to buy the product. In Figure 11.1, we can see that a consumption of Q1 units of the product results in a consumer surplus as indicated by the area enclosed by the triangle X, Y, Z. The producer surplus is indicated by the area enclosed by the triangle A, B, C.

Figure 11.2 considers the cost to society at large that results from the production process, and indicates that the cost of the same level of consumption (Q2) is now higher. In this case, the supply curve (marginal pollution cost + marginal supply cost) is steeper, so for the same quantity of the good the consumer would pay a higher price, reflecting the cost

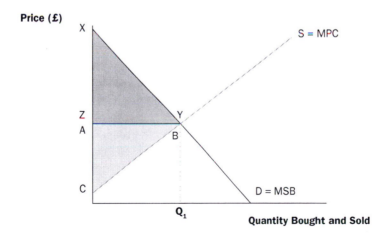

Figure 11.1 Quantity of products bought and sold determined by supply and demand curves

Source: Drawn by Imogen Shaw

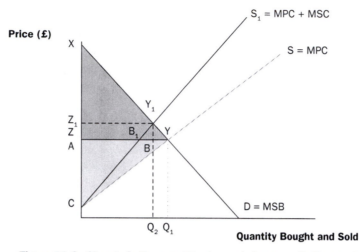

Figure 11.2 *Change in the quantity demanded once pollution costs are internalized*

Source: Drawn by Imogen Shaw

of production including the negative impact of the pollution. The consumer surplus is now less, indicated by the smaller area enclosed by the triangle X, Y_1, Z_1, and similarly the producer surplus has been reduced by internalizing the cost of pollution control (triangle C, A, B_1).

The main problem with this approach is that its conclusion is that too much of the *good* is produced, rather than suggesting that either the good should be produced in a way that generates less pollution, or that the good should not be produced at all. All this sort of economic analysis can do is to predict the 'equilibrium' point where the optimal amount of production occurs. If it were to result in concrete policy, it would also require that the full costs of pollution can be accurately measured, whereas in the case of sophisticated pollutants that might cause many different types of environmental damage over a long period of time, this is impossible to achieve in practice.

The neoclassical response to the problem of pollution might be to introduce a Pigovian tax (see Section 3.1) to impute the true cost of production into the price paid by producers – what a neoclassical economist would call 'internalizing the externality'. An alternative policy would be to introduce a requirement for a polluter to buy a permit to produce the pollution. These permits might then be traded. Factories whose cost of reducing pollution was lower might choose to cut their pollution and sell their spare permits to companies who would find it more difficult – and therefore more expensive – to reduce their own emissions. This would be an efficient market outcome, and an efficient way of achieving the desired reduction in polluting emissions. Section 3.5 describes how such a policy has worked in the case of sulphur dioxide pollution in the USA; Section 13.3 describes how such a policy might work as a way of reducing the CO_2 emissions that are causing climate change.

11.2. Negotiating shared political action

These market solutions to the problem of pollution arise from an approach to social life that prioritizes the economy, and seeks solutions to social problems in the market. While pollution is defined as an example of market failure, market solutions are nonetheless frequently proposed. However, there are alternative approaches that seek solutions in the arena of politics rather than economics. Those producing pollution can be prevented from doing so either by banning the pollution and creating legal sanctions against those who continue to pollute, or by limiting the amount of permissible pollution, or by regulating the way in which pollutants can be allowed into the environment.

Acid rain is frequently cited as an example of a pollution problem that was successfully contained by a trading scheme in the USA, while the European Union took a regulatory route.

Acid rain occurs when pollution gases from the burning of coal and oil – sulphur dioxide (SO_2), sulphur trioxide (SO_3) and nitrogen dioxide (NO_2) – come into contact with water vapour, which turns them into sulphuric and nitric acids, which are damaging to plant and animal life. Because the pollution travels across national borders, an international agreement is the only way of tackling the problem. In the early 1980s, the Scandinavian countries (being significant net recipients of pollution) were instrumental in organizing international conferences to attempt to reach an agreement, but these moves were blocked by the UK (a significant net emitter). In 1988, the EU passed the Large Combustion Plant Directive, which set targets for reductions by the Community's major polluters, which were then tightened in 2001.

Further research conducted by the EU indicated that another type of airborne pollutant – particulate matter, i.e. very small particles of soot from the burning of fossil fuels – were causing serious disease amongst European citizens and reducing life expectancy by eight months on average and two years in the worst affected areas (European Commission, 2005). The EU had signed a Convention with the USA and Canada in 1979; this was later signed by another 50 countries, and has led to protocols limiting emissions of all the main airborne pollutants. In 2005, the Commission published a strategy to reduce emissions further by 2020. The strategy included 'a mix of policy measures, including new or revised legislation, access to EU funding and international cooperation through the Convention on Long-Range Transboundary Air Pollution' (European Commission, 2005: 3). The legislation covers a range of airborne pollutants, including exhaust emissions of particulates and emissions from all sources of gases, including sulphur dioxide, nitrogen oxide, ammonia and volatile organic compounds.

Figure 11.3 indicates the success that strong legislative action has had on reducing the levels of air pollution in the EU, in this case in terms of

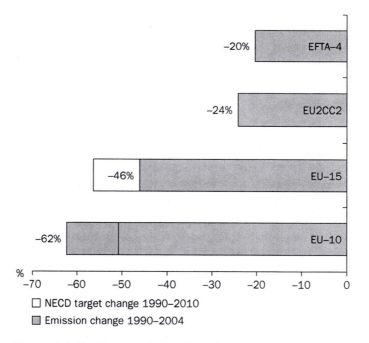

□ NECD target change 1990–2010
▨ Emission change 1990–2004

Figure 11.3 *Percentage changes in emissions of acidifying substances (SO2, NOX and NH3) over the period 1990–2004, and comparison with NEC Directive targets*

Source: Data from the European Environment Agency (2007), Fig. 4.5

acidifying emissions. Overall, this is an encouraging indication of what regulatory action can achieve, although there are also areas of concern. Emissions of the gases that are responsible for acid rain have decreased by 46 per cent in the 15 countries that were members prior to the enlargement between 1990 and 2004, although those of Spain, Greece and Portugal have all increased. This has been mainly due to a shift from coal-fired to gas-fired power stations, the introduction of technologies to remove sulphur dioxide emission from power station emissions, and the decline in heavy industry in the former East Germany. Declines in industrial emissions of nitrogen oxides have been balanced out by increases in emission from cars and trucks (EEA, 2007). This indicates that it is the nature of our consumption behaviour and lifestyles that is driving the pollution trends, an issue we turn to in the following two sections.

First, we will briefly consider a positive example of international cooperation in the case of one specific environmental pollutant: the emission of gases that caused a hole in the earth's protective ozone layer. The ozone layer is vital for the survival of life on earth, because it protects plants and animals from ultraviolet radiation. A type of gas that was used for fridge coolants and aerosol propellants known as chlorofluorocarbons (CFCs) was discovered to be causing chemical changes in the ozone layer, resulting in thinning and even the absence of ozone for some parts of the year, especially in the northern hemisphere. Without the protection of the ozone layer, dangerous levels of UVB radiation are able to penetrate to the earth's surface, causing skin cancers and potentially blindness in humans and animals, and destroying plant life, including agricultural crops.

International negotiations to tackle the emissions of CFCs were initiated by the UN Environment Programme (UNEP) in 1976, and by 1985 the Vienna Convention encouraging cooperation and exchange of research information was agreed, although it did not include any legally binding controls of targets. The research that followed indicated the existence of a hole over the Antarctic, and serious thinning of the ozone layer over the Arctic. The Montreal Protocol on Substances that Deplete the Ozone Layer was adopted in September 1987; it consisted of a series of legally binding phase-out schedules for the various gases concerned. It has been amended several times since the original protocol, most recently again in Montreal in 2007.

Figure 11.4 indicates the impact of the Montreal Protocol on the production of ozone-depleting gases. According to UNEP, 'In 1986 the

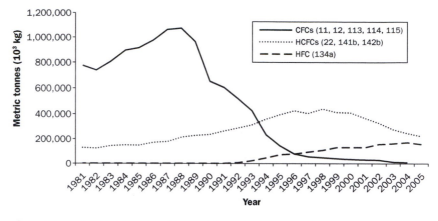

Figure 11.4 *Worldwide production of CFCs, HCFCs and HFCs*

Source: Data from the European Environment Agency website based on data derived from the
UNEP Ozone Secretariat

total consumption of CFCs world-wide was about 1.1 million ODP
tonnes; by 2006 this had come down to about 35,000 tonnes' (UNEP,
2008: 6). In parallel the production of the much less destructive HFCs has
increased. Former UN Secretary-General Kofi Annan considered that
'Perhaps the single most successful international agreement to date has
been the Montreal Protocol'. UNEP's suggestions as to the reasons for
this success are listed in Box 11.1. We can hope that these guidance notes
will be of use to those negotiating treaties to tackle the much more serious
and urgent issue of greenhouse gas emissions, but it is important to
remember that alternatives to CFCs could be found that would not
threaten the functioning of industries or the lifestyles of consumers. This
sort of simple, technological switch will not be possible in the case of
climate change, which means the problem is different in magnitude and
nature (see Chapter 13).

Box. 11.1

Lessons from the Montreal Protocol that could be applied to other environmental issues

- Adhere to the **precautionary principle** because waiting for complete
  scientific proof can delay action to the point where the damage has become
  irreversible.

- Send consistent and credible signals to industry (e.g. by adopting legally binding phase-out schedules) so that it has an incentive to develop new and cost-effective technologies.

- Ensure that improved scientific understanding can be incorporated quickly into decisions about strengthening (or weakening) the provisions of a treaty.

- Promote universal participation by recognizing the 'common and differentiated responsibility' of developing and developed countries and ensuring the necessary financial and technological support to developing countries.

- Control measures should be based on an integrated assessment of science, economics and technology.

Source: UNEP (2008)

11.3. Working with nature to minimize pollution

As was noted at the beginning of this chapter, the natural environment can break down all wastes, given sufficient time, because it has what an ecologist would define as an 'assimilative capacity', i.e. it has the ability to absorb pollutants. But there are some important points to note about the environment's assimilative capacity:

- It is limited – we should use it as a scarce resource rather than seeing it as a bottomless pit.
- It depends on the flexibility of the ecosystem and the nature of the waste. Some wastes are more degradable than others, and some pollution is what is defined as 'persistent' meaning that it stays in the environment for a very long time – radioactive pollution and the type of chemicals known as PCBs are examples.
- Its ability to degrade waste depends on the rate at which the waste is discharged: pollution reduces the capacity of an environmental medium to withstand further pollution.

An ecologically minded economist would take a **holistic** view of pollution: we can view nature as a whole system, of which we are a part. Within such a view, all parts of the system interact, and pollution we put into the environment will stay with us since we are a closed system. Therefore it does not make a great deal of sense to put anything into our environment that is toxic or harmful to life, since we are living systems and it is likely to impinge on us at some point in the future.

Rather we should work with the system of nature, respecting what we have learned from scientific study about how natural systems work – not overloading them – and minimizing the harmful substances we produce.

Permaculture is a way of looking at the world as a system and 'creating sustainable human habitats by following nature's patterns' (Holmgren, 2002: 2). It follows a series of principles that its proponents claim arise from nature, such as use and value diversity, produce no waste, and use small and slow solutions. It challenges the immediate rush to action that is the response of many to first learning about the environmental crisis; instead it cautions that we should engage in maximum contemplation and minimum action. 'Permaculture is not a set of rules; it is a process of design based around principles found in the natural world, of co-operation and mutually beneficial relationships, and translating these principles into actions.'

If we apply the thinking of permaculture to production systems we arrive at an approach known as 'industrial ecology'. In this economic paradigm we have a real circular flow, in contrast to the limited 'circular flow' of neoclassical economics (see Figure 11.5). The conception of the economy is as a limited system, within which all waste products must become useful inputs to the next production process, and design focuses on the achievement of 'closed loops' where neither materials nor energy are wasted:

> Industrial ecology provides a powerful prism through which to examine the impact of industry and technology and associated changes in society and the economy on the biophysical environment. It examines local, regional and global uses and flows of materials and energy in products, processes, industrial sectors and economies and focuses on the potential role of industry in reducing environmental burdens throughout the product life cycle.
>
> (International Society for Industrial Ecology website:
> http://www.is4ie.org/, accessed 13 September 2010)

While this may seem rather abstract, we probably all have experience of a process which operates in this way – the production of compost from kitchen waste. The potato peelings and grass cuttings that we put into our compost heaps biodegrade naturally and become new soil, which can then be used as the basis for the growth of more food. Hence we have a perfect, closed loop of production → consumption → waste → natural decomposition → new production. These are the sorts of cycles that

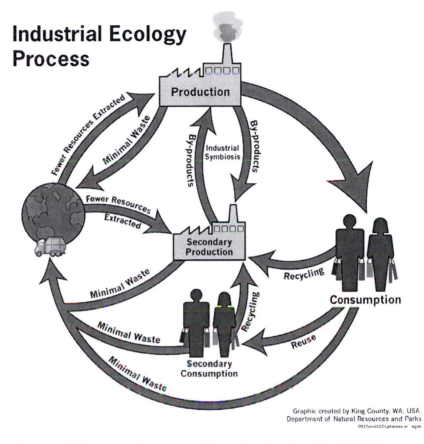

Industrial Ecology Process

Graphic created by King County, WA, USA,
Department of Natural Resources and Parks

Figure 11.5 *The production–consumption process as viewed by industrial ecology*

Source: Drawn by Wendy Gable Collins; my thanks to the International Society for Industrial Ecology for permission to reproduce this graphic free of charge

industrial ecologists seek in industrial production processes (see the principles in Box 11.2). At the industrial level, we can think of the use of natural materials such as wood that absorb carbon dioxide while they are growing and fix it until they decay – rather than bricks, which have a huge amount of **embodied energy** because of the fossil fuels that are used to fire them. Thus, in an ecologically benign industrial system, pollution would be minimized or eliminated altogether simply by designing 'with nature in mind'.

Box 11.2

Principles of production to match the metabolism of the natural world

- Buildings that, like trees, produce more energy than they consume and purify their own waste water;
- Factories that produce effluents that can be used as drinking water;
- Products that, when their useful life is over, do not become useless waste but can be tossed on to the ground to decompose and become food for plants and animals, and nutrients for soil; or, alternatively, that can be returned to industrial cycles to supply high-quality raw materials for new products;
- Transportation that improves the quality of life while delivering goods and services; and
- A world of abundance, not one of limits, pollution and waste.

Source: Porritt (2005)

11.4. One man's meat is another man's poison

An orthodox economist would make two fundamental assumptions about pollution: there is agreement about what it is, and it is an inevitable by-product of industrial processes. The previous section began to explore whether we might be able to tackle pollution more creatively if, rather than seeing it as something that needs to be dumped, we can begin to see it as something that can be assimilated naturally into the environment, and we work to develop industrial processes that facilitate this. But we might go even further, and begin to ask whether the concept of pollution is actually one that is relative to time and place.

In 1966, anthropologist Mary Douglas famously defined 'dirt' as 'matter out of place', a definition which can help to guide our thinking about what pollution is. Our definitions of what is 'unclean' are culturally relative, and change over time and between different societies. Sometimes the change is the result of progress in scientific knowledge and understanding, so that the poster shown in Figure 11.6 seems anachronistic now that we know the health dangers from industrial pollution. Perhaps one day nuclear power plants will seem similarly anachronistic once the health effects of radionuclide contamination are

Figure 11.6 *'The Smoke of Chimneys is the Breath of Soviet Russia'*

Source: Reproduced freely thanks to Wikimedia commons

better understood. Douglas argued that what we consider to be pollution has as much to do with our attitude towards order, control and risk as it does to do with scientific understanding: 'some pollutions are used as analogies for expressing a general view of the social order' (Douglas, 1966: 3).

Air pollution, which has formed the focus of attention for this chapter, is a useful example to trace the development of ideas about pollution. London was the world's first mega-city, and the first to experience serious pollution from coal-burning – as early as the Middle Ages. The most serious pollution event occurred in 1952, when the 'Great Smog' killed more than 4,000 people between 5 and 8 December (Mayor of London, 2002). The smog was caused by the burning of coal in domestic fires and in industrial premises and power stations. The UK Clean Air Act, passed in 1956, required the siting of power stations in rural areas and introduced a system of 'smokeless zones' in which only special smokeless fuel could be burnt. The 1968 Clean Air Act required the use of tall chimneys for premises that produced noxious fumes. Ultimately, concern with air pollution led to a switch to the burning of natural gas for cooking and heating, and to the increase in gas-powered and nuclear-powered electricity-generating stations.

The last paragraph indicates that pollution was considered to be something that could be seen or smelt, and the damage from which was evident locally. Once pollution could be moved further afield or made invisible, it was 'out of sight and out of mind'. Hence, nuclear power was defined as 'clean energy', because it could be generated far from the centres of population, and because its health consequences could not be directly traced to it and might occur many years later or many miles away. It is evident now, of course, that by shifting from burning coal to other fossil fuels and nuclear energy, we have simply increased our reliance on technologies that are causing radioactive contamination and climate change. The mindset that considers pollution as an **externality** has thus entrenched an economy and production-and-consumption pattern that is out of kilter with nature. Extending Kenneth Boulding's image of the spaceship earth (see Section 2.1), we can see that we have continued to view the environment from the perspective of a cowboy rather than a spaceman, merely shifting the pollution further from where we live, rather than actually recognizing that it is still within our ecosystem and that therefore we must minimize it.

11.5. Case study: The Athabasca tar sands as an example of water pollution

As we have used the air as a free sink for pollution, so we have used waterways to wash away the wastes from our consumption and especially our production systems. Hanley et al. (2001) link the pollution of Scotland's waterways to increasing population and industrialization from the beginning of the nineteenth century onwards. They identify the five main causes of the pollution of watercourses as: inadequately treated human sewage; drainage of contaminated surface water (for example by oil leaks or factory waste); industrial discharges; discharges from abandoned mines; and agricultural effluent. The pattern is the same in the former Eastern European economies and the developing economies of the South. Pollution reaches a peak following initial industrialization, and is then reduced as a result of public and political opposition, which leads to legislative controls.

> The area near Fort McMurray and the Athabasca River, Alberta, in Canada is the scene of the latest rush for black gold. The world's biggest oil companies – Shell, Chevron, Exxon, Total, Occidental and Imperial – are investing heavily in extracting usable oil from the heavy bitumen deposits found in the shale sands of the area. They are mining 1.3 million barrels of oil per day, and expansion is proceeding rapidly: by 2050, Canada could be the second-largest oil producer in the world. The consequence has been massive contamination of the water table and hundreds of square kilometres of toxic waste ponds: 'So far, nearly 180 sq miles (470 sq km) of forest have been felled by tar sands miners and giant lakes of toxic waste water cover a further 130 sq km'
>
> (Vidal, 2008).

The tar sands are an example of what neoclassical economists would consider a technological advance to outweigh the depletion of a non-renewable resource, in this case lighter crude oils that can be drilled in the Middle East and Russia. But the environmental cost of extracting the oil is heavy, not only in terms of water pollution but also because the extraction is itself intensive in terms of carbon dioxide emissions. The tar sands also raise questions about environmental justice, since the area is home to five groups of Canadian first nations people – the indigenous inhabitants of North America, who have lived in harmony with their ecosystem for thousands of years. It is these people who are suffering the cancers and birth defects as a result of the oil extraction.

Summary questions

- What problems would you face as Minister for the Environment in setting the rate of a Pigouvian tax on a form of industrial effluent that is toxic to fish?
- Why do you think that the Montreal Protocol was effective?
- What is the link between industrial ecology and the green economists' theory of the closed-loop economy?

Discussion questions

- How would you define pollution? How can you decide what level is acceptable?
- In what ways is the Montreal Protocol relevant, or not relevant, to tackling the problem of climate change?
- Can the design principles of industrial ecology solve the pollution problem, or is there just too much production?

Further reading

Hanley, N., Shogren, J. F. and White, B. (2001), *Introduction to Environmental Economics* (Oxford: Oxford University Press), ch. 11 on water pollution: a useful case study of one type of pollution.

Nagle, J. C. (2008), 'The Idea of Pollution', *Notre Dame Legal Studies Paper No. 07–05* (Notre Dame, IN: Notre Dame Law School): presents a philosophical and legal discussion of what we mean by pollution.

UNEP (United Nations Environment Programme, 2008), *Basic Facts and Data on the Science and Politics of Ozone Protection* (Geneva: UNEP): a scientific account of one self-contained environmental problem that was successfully resolved by international negotiation.

12 Globalization vs. localization

Since the end of the Second World War, and in a pattern of accelerating speed and intensity, the economies of the world have become more intertwined and interdependent, in a process known as 'globalization'. The consequences have changed every aspect of life for the citizens of the richer, Western nations – and for the poorer South even more. Some of the changes have been beneficial to individuals, including rapid increases in monetary incomes, a wider range of consumer goods, facilitation of international travel and communication links. For corporations that have a global base, as opposed to the small businesses that make up our local economies, these changes have offered many advantages, including access to workforces who demand lower wages, the ability to choose to pay tax in countries with more favourable regimes, and reduced costs of marketing via the establishment of global brands. This chapter attempts to balance the advantages and disadvantages of globalization, from the perspective of the environment, as well as the world's people.

Section 12.1 offers evidence that globalization has, indeed, been the dominant process of change in the world economy in recent years, and presents the views of some of those who welcome this development. Section 12.2 presents the more critical views of some of the doubters – those who are drawing attention to the negative consequences of this trend, highlighting the losses and who the primary losers are, and particularly questioning the unequal power relationships on which globalization depends. Section 12.3 presents the case of those who oppose globalization and argue for a system of localization, with the world returning to an interconnected network of strong local economies. It offers a view of provisioning based on naturally defined units, with trade playing a residual role of providing only those goods that cannot be produced within these 'bioregions'. Finally, Section 12.4 presents a case study exploring the power relationships at play in the global trade system, through a short history of the Banana Wars.

12.1. All aboard the globalization rollercoaster

What exactly do we mean by 'globalization'? The following quotation from Saleh Nsouli, European Director of the IMF, includes many of the most obvious features:

> 'Globalization' can mean something different to each of us, so let me specify what I have in mind: the increasing integration of economies in the world, particularly through the international flow of goods, services, and capital – and, increasingly, people (labor) and knowledge (technology and information). Economic globalization is a historical process, driven mainly by invention and innovation, as well as economic policy.
>
> (Nsouli, 2008)

So globalization is a process which eases the movement of goods and services – and the money that is invested to produce these and made from the sale of them – between the different countries that make up the world community. Nsouli makes it clear that the freeing-up of markets in this way has not yet extended to either the movement of people, because most countries maintain strict immigration rules, or technology, because a technological advantage is likely to be protected by the company that owns it. Hence we already see that there are limits to the freedom that globalization is said to bring, and that some groups have more power in the process than others.

Mr Nsouli continued his speech on behalf of the United Nations by listing some of the changes that the process of globalization has brought, including the increase in the ratio of trade to global **GDP** as a whole of 42 per cent between 1980 and 2007, and an increase in the proportion of this that is funded by foreign direct investment, from 6 to 32 per cent over the same period. In other words, more of the economic activity that takes place in the world today is funded by a country other than the one in which the productive activity is based, and an increasing proportion of it is eventually sold outside the country in which it is produced. Indeed, this is a stark demonstration of the rapid shift that globalization has brought, but there is no particular reason to think that this is a beneficial process. The production of goods in countries where wages are lower results in lower prices for the consumers in countries that buy these products (largely the developed Western economies), but does it benefit the poorer countries where more of the world's production now takes place? We will examine the answer to this question in a later section.

Table 12.1 presents data indicating the consequences of globalization for seven developed and developing countries. While all have seen an increase in their level of imports, it has been particularly rapid for China, India and Mexico. The figures for energy use make clear the close link between economic development via a globalization process and the use of energy, and hence its relationship with climate change. The data for mobile phone use are included to give an indication of the way in which communications are extending – a process that is very recent and also includes internet communications, for which a similar trend can be observed in the data. Finally, the most striking data are those relating to the enormous expansion in the flows of investment into the world's poorer countries – again especially India and China. This supports the arguments of some of the critics of globalization – that it is dominated by the movement of capital.

Supporters of globalization argue that the freeing-up of international markets permits the most efficient use of resources, since they will travel to the countries where they can be exploited most efficiently. They also tend to revel in the homogenization of world culture – the elimination of boundaries and differences between countries. A leading populist example is Thomas Friedman, whose book *The World Is Flat* (2005) won the Pulitzer Prize. Box 12.1 lists what Friedman calls the 'ten flatteners' – the ten forces that have led to a world of sameness rather than difference, a process which he celebrates. It is clear from Friedman's list, however, that his interest is personal rather than political: he makes no mention of the freeing of international capital which underpins globalization as an economic process. Friedman is excited about the spread of his culture – US capitalist culture – to all corners of the globe, so that he and his colleagues can play golf in Bangalore with others who speak their language and are willing to accept their currency.

Table 12.1 *The rapid process of globalization in figures*

	Bangladesh	China	Germany	India	Kenya	Mexico	UK
Imports[a]							
1980	18	11	25	9	36	13	25
1990	14	16	25	9	31	20	26
2000	19	21	33	14	32	33	30
2008	28	28	40*	30	39	30	29*
FDI, net inflows[b]							
1980	n/a	570.0	342.4	79.2	79.0	2,090.0	10,122.8
1990	3.2	3,487.0	3003.9	236.7	57.1	2,549.0	33,503.7
2000	280.4	38,399.3	210,085.4	3,584.2	110.9	17,941.9	122,156.8
2007	652.8	138,413.2	51,543.3	22,950	727.7	24,686.4	197,766.2
Mobile phones[c]							
1980	0	0	0	0	0	0	0
1990	0	0	0	0	0	0	2
2000	0	7	59	0	0	14	74
2008	28	48	131	30	42	71	123
Energy use[d]							
1980	94	610	4,597	304	486	1,429	3,575
1990	111	760	4,477	377	479	1,478	3,708
2000	133	876	4,174	453	481	1,533	3,971
2006	161	1,433	4,231	510	491	1,702	3,814

Notes: asterisked data are for 2006

[a]as a percentage of GDP   [b]Foreign Direct Investment: Current US$ million   [c]per 100 people   [d]kg of oil equivalent per capita

Source: Data from the World Bank's World Development Indicators database

---

Box 12.1

Ten forces that flattened the world

1. The new age of creativity: when the walls came down and the windows went up: *the end of communism removed an ideological competition with global capitalism*
2. The new age of connectivity: when the web went around and netscape went public
3. Work-flow software
4. Uploading: harnessing the power of communities
5. Outsourcing: *subcontracting work to employees outside the company who may have different terms & conditions and wage-rates*
6. Offshoring: *subcontracting work to employees in other countries, where employment legislation may be different*
7. Supply-chaining: *extending the distance between producer and consumer*
8. Insourcing: *increasing the level of specialization within firms*
9. In-forming: *using information from the internet to gain competitive advantage*
10. Steroids: digital, mobile, personal and virtual: *the influence of a range of personal consumer electronics gadgets making information available anywhere, any time*

Source: Friedman (2005) with author's interpretations in italics

---

12.2. Global sceptics

Now that the market is global, neoclassical economics would suggest that this freely functioning market, like any other, will achieve equal outcomes for all participants. Some may be further ahead on the curve towards growth and development, but the process of 'growth convergence' (Barro and Sala-í-Martin, 2004) will mean that, over time, those who are further behind will catch up, meaning that the process of globalization will offer significant benefits to the poorer nations of the world. Within this economic paradigm, any evidence of uneven economic development would be an example of market failure. Table 12.2 presents evidence that, although most of the low-income countries listed have achieved rapid rates of growth, these have varied greatly between nations.

One of the most prominent criticisms of untrammelled growth has come from an orthodox economist who was formerly a leading proponent of globalization and Chief Economist of the World Bank,

Table 12.2 Varying rates of growth in different low-income countries, 1997–2007

GDP per capita growth	Countries
More than 50%	Chad, Cambodia, Myanmar, Mozambique, Nigeria, Sierra Leone, Tajikistan, Vietnam
25–50%	Bangladesh, Burkina Faso, Ethiopia, The Gambia, Ghana, Krygyz Republic, Laos, Madagascar, Mali, Nepal, Pakistan, Saō Tomé & Principe, Tanzania, Uzbekistan
0–24%	Benin, Kenya, Guinea, Malawi, Mauritania, Niger, Rwanda, Senegal, Uganda, Republic of Yemen, Zambia
Less than 0%	Burundi, Central African Republic, Comoros, Dem. Rep. of Congo, Côte D'Ivoire, Etritrea, Guinea-Bissau, Liberia, Papua New Guinea, Solomon Islands, Togo, Zimbabwe

Source: Ahmed (2009), based on data from the IMF and the World Bank

Joseph Stiglitz. Stiglitz's primary concern is that, although incomes have risen in many of the world's poorer countries, within and between those countries there has been a parallel and rapid rise in inequality. There is a great deal of data from the World Bank (see Figure 12.1) to underpin this case:

> Over the past two decades, income inequality has risen in most regions and countries. At the same time, per capita incomes have risen across virtually all regions for even the poorest segments of population, indicating that the poor are better off in an absolute sense during this phase of globalization, although incomes for the relatively well off have increased at a faster pace. Consumption data from groups of developing countries reveal the striking inequality that exists between the richest and the poorest in populations across different regions.
>
> (IMF, 2008)

In addition, Stiglitz points out that smaller countries with weaker currencies that become rapidly exposed to competition in the global economy are vulnerable to the movement of international capital out of their economy, and the destructive consequences of this.

The critique of globalization, which is now shared to some extent by employees of the World Bank, originated amongst economists based in the countries of the South who were not controlling the economic changes

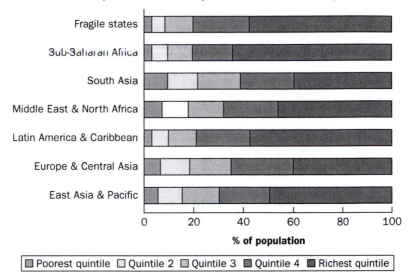

Figure 12.1 *Share of the poorest and richest quintiles in national consumption in different regions of the world*

Source: World Bank

affecting their countries. As one of the founders of the International Forum on Globalization, Vandana Shiva has long been a prominent critic of the consequences of globalization for the world's poor, and identified the negative consequences for local environments in India as well as the global climate from the energy-hungry global system. Martin Khor, of the Third World Network based in Penang, Malaysia, identifies a range of factors that have led to disillusionment amongst policy-makers in the South:

> [T]he lack of tangible benefits to most developing countries from opening their economies . . . the economic losses and social dislocation that are being caused to many developing countries by rapid financial and trade liberalization; the growing inequalities of wealth and opportunities arising from globalization; and the perception that environmental, social and cultural problems have been made worse by the workings of the global free-market economy.
> (Khor, 2001b: 1)

The system that has become globalized is a particular economic system: corporate capitalism. Thus it is unsurprising that the factor of production which has travelled the world most freely is capital. The freeing of capital from any degree of political or democratic control in the 1980s created a situation where money dominated at the expense of labour, which no longer had the right to control the movement of capital through the governments it elected. This delinking of political and financial systems has had severe consequences for those outside the charmed circle of countries with powerful currencies, and has also led to the domination of the real economy by finance capital (Mellor, 2010). It has been estimated that at the turn of the millennium, 97 per cent of financial transactions that took place in the global economy had no contact with the real economy at all (Khor, 2001b). Most money is made through speculative investment in other markets and derivative markets, the so-called 'casino economy'.

It is fairly obvious, although the point is rarely made by either supporters or critics of globalization, that a globalized economy is an energy-intensive economy. At the hard end, globalization is about increasingly lengthy supply chains between producer markets (where wages and environmental standards are low) and consumer markets (where wages are high, meaning that people can afford to buy products and their own environments are better protected). The movement of goods from the one to the other – for example, from China to the UK – relies on the combustion of petroleum in one form or another, and therefore generates carbon dioxide emissions. Globalization also tends to increase demand for products, especially in poorer countries that aspire to the lifestyle of those they export to, which again increases demand for energy. At the softer end, the connections we make when we travel, or relationships we start initially through virtual, internet connections, create demand for more travel for us to meet our far-flung friends, what George Monbiot has referred to as 'love miles'.

12.3. From globalization to the bioregional economy

Proposition is always a more powerful move than opposition, and a significant proposal for relocalizing economics was made by Colin Hines in his 'manifesto' for localization published in 2000. He offers his own, highly critical, definition of globalization:

> **Globalization** n. 1. the process by which governments sign away the rights of their citizens in favour of speculative investors and transnational corporations. 2. The erosion of wages, social welfare standards and environmental regulations for the sake of international

trade. 3. the imposition world-wide of a consumer monoculture. Widely but falsely believed to be irreversible – See also financial meltdown, casino economy, Third World debt and race to the bottom (16th century: from colonialism, via development).

(Hines, 2000: 4)

As the quotation makes clear, already in 2000 critics of globalization were suggesting that the free movement of capital and the speculation it gave rise to were a threat to financial stability. It was from these critics that the alarm was sounded about the vulnerability of the financial system – nearly a decade before the credit crunch of 2008/9.

Other critics were more concerned about the consequences for those in the developing world, who were being brought into the global economy by their own elites and by pressure from the more powerful trading nations via the World Trade Organisation (WTO). How would they fare in a competition with countries whose economic power was so much greater than theirs? Some answers can be found in work by Lines (2008), who concludes that trade is actually 'making poverty'. His analysis of trade figures (see the data presented in Table 12.3) makes it clear that only a small proportion of countries are actually benefiting, and within those countries the benefits accrue to those who control the trade, not those who work to grow or produce the export goods. In the case of fruit and vegetables, for example, he found that two-thirds of exports from poorer countries are produced by just eight countries, and that Keyna is the only country to have benefited from the increase in the quantity of vegetables exported from sub-Saharan Africa. The reason for this is the

Table 12.3 Changes in the terms of trade of some country groups, 1980–2 to 2001–3

Group	Annual average 1980–2	Annual average 2001–3	% change
Developed economies	95.7	103.3	+7.9
Developing economies	117.3	97.7	−16.7
Developing economies: Africa	131.7	100.0	−24.1
Least developed countries	144.0	93.3	−35.2
Landlocked countries	114.7	96.3	−16.0
Sub-Saharan Africa	124.0	98.3	−20.7

Source: Data from UNCTAD; calculations in Lines (2008); table reproduced from Cato (2008)

'terms of trade', i.e. the prices that are paid for these on world markets. In spite of the rhetoric about 'free trade', in fact the markets are dominated by the wealthy importer countries (see Box 12.2).

The consumption of fossil fuels to drive the globalized economy is one side of the relationship between globalization and the environment. The other side is the increased lack of security in terms of basic resources that results from lengthy supply chains. Figure 12.2 illustrates the food surpluses or deficits of a range of countries, and indicates that the UK is especially vulnerable, having relied on the sale of financial services to pay for its food imports. Relying on international markets for your supply of basic food commodities may seem like a reasonable strategy in times when food is plentiful and cheap, and when your currency is strong, but climate change is likely to make global food harvests unpredictable, leading to higher and more volatile prices.

The supporters of localization have achieved considerable rhetorical advantage through their critique of the globalized economy – especially its impact on the planet. Where they have been less successful is in defining the local economy. A concept that can help to ground this idea of localization, and to ensure that the local economies that develop are closely embedded in natural systems, is that of the 'bioregion'. Bioregionalism is a culture of living that acknowledges ecological limits (McGinnis, 1990). From an economic point of view, the bioregion could be used as an area that could aim to be largely self-sufficient in terms of its basic resources, such as food, energy, water, and products such as clothes and furniture (Cato, 2007). It should also take responsibility for

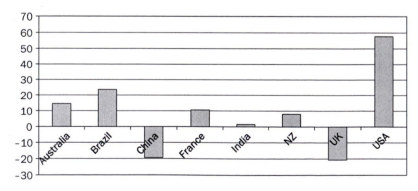

Figure 12.2 *Food surpluses and deficits for a selection of developed economies (2005; m tonnes)*

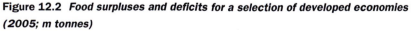

Source: Author's graphic based on FAO data

Box 12.2

Tesco and the conflict between trade and the environment

As the issue of climate change has risen up the political agenda, the link between carbon dioxide emissions and globalization has become a source of increasing debate. Since the corporations who dominate the globalized economy rely on lengthy supply chains for their profits, they have been some of the first to enter the debating arena. Tesco, for example, has funded academics at Oxford University's Environmental Change Unit to help them reduce the carbon impact of their supply chain. At Manchester University, they have gone even further, investing £25 million in a Sustainable Consumption Institute, which is founded on the premise that the solution to the environmental crisis is not to limit consumption but to consume more intelligently (Attwood, 2007).

In a speech he made in 2007, Terry Leahy, the then CEO of Tesco said:

> We cannot avoid the fact that transporting a product by air results in far higher carbon emissions than any other form of transport. We are not willing to avoid the hard fact that there is a conflict between the issue of carbon emissions and the needs of some of the poorest people on earth whose lives are improved by the ability to sell in our markets products which are brought here by air. There is a strong international development case for trading with developing countries. So, the question is: should we shun Fairtrade horticulture from East Africa to save CO2, or champion it as an important contribution to alleviating poverty?
>
> (Leahy, 2007)

Tesco seeks to portray its business model as motivated by a desire to alleviate poverty rather than boost corporate profits – although, as has already been discussed, recent empirical work suggests that the terms of trade limit the extent to which producing food for Western markets can help the world's poorest people. The shift to a low-carbon economy needs all sectors of society to be involved; one should not be too cynical about the motives of corporate investors, since their survival will require them to follow this trend. However, one might be concerned at the huge level of investment they are able to make in research, and wonder how likely their researchers are to discover that local, self-reliant food systems are better for the environment.

its own waste, thus becoming a closed-loop economy, which green economists argue is the most energy- and resource-efficient type. This is therefore a highly geographical approach to providing basic resources (or 'provisioning'), as distinct from the boundaryless approach that is favoured by globalizers.

Sale provides principles to guide a bioregional economy (reproduced in Box 12.3): it is clear that this is a locally based provisioning economy rather than a consumer economy. In material terms, it focuses on needs rather than desires; bioregional citizens achieve their well-being through relationship and self-expression rather than material consumption. It is also clear that, like the economies that dominated Europe before the emergence of capitalism, it is an economy with a clear moral and social motivation.

So how much trade is compatible with a responsible climate policy? Are we to contemplate a bleak future where we spend the winter living on turnip soup and have no friends outside our own town or city? Conventional economists have a barely concealed dread for this sort of rigorously self-sufficient approach, which they have labelled as 'autarky'. A moderate alternative to such a bleak picture is the idea of 'trade

---

Box 12.3

Kirkpatrick Sale's essential elements to guide a bioregional economy

1. All production of goods or services would be based primarily on a reverence for life.
2. The primary unit of production would be the self-sufficient community, within a self-regarding bioregion, which would strive to produce all its needs.
3. Consumption would be limited, for it is not a rightful end in itself but merely a means to human well-being.
4. Everything produced and the means of its production would embody the four cardinal principles of smaller, simpler, cheaper, safer.
5. The only jobs would be those that enhance the worker, contribute to the immediate community, and produce nothing but needed goods.
6. All people who wish to do so would work.
7. All economic decisions would be made in accordance with the Buddhist principle: 'Cease to do evil; try to do good.'

Source: Sale (2006b)

subsidiarity', which proposes that, by analogy with the political subsidiarity, goods and services should be provided at the lowest appropriate level. The aim would be self-reliance – sufficiency in basic commodities and especially staple foods, with trade returning to its role in providing luxuries out of season, and crops that cannot be produced because of limitations of expertise or climate. This would offer significant advantages in terms of local identification and stronger embedding in a local environment (Morgan et al., 2006), as well as reducing the energy-intensity of our food production and distribution system.

12.4. Case study: The Banana Wars

The UK fruit market provides an interesting case study of the struggle between globalization and localization in the supply of a basic food product. Traditionally, British consumers bought mainly apples and pears, which were in season in the autumn and could last for many months if stored and kept cool. A range of varieties was grown that fruited and ripened through the late summer and autumn months, extending the period during which fresh fruit was available. In addition, citrus fruits from southern Europe were imported as speciality foods, and the Christmas period was marked by the arrival of silver-wrapped tangerines. However, assiduous marketing and the availability of refrigerated container shipping led to the growing popularity of the banana, and in 1998 it overtook the apple as the fruit most frequently consumed in the UK. The energy consequences of this consumption shift are clear, both in terms of the refrigeration of the ships and fuel oil burned to transport the fruit from the tropics to Western markets.

The global banana market is dominated by three major producers: Chiquita International, Dole Food and Fresh Del Monte Produce, who between them supply about 56 per cent of the world's bananas (this figure and much of the information on which this case study is based is taken from Robinson, 2009). This is a large market, estimated to be worth approximately US$5 billion per year, with something like 6.5 million tonnes of bananas being bought and sold. £575 million worth of them (or 7 billion per year) being eaten in Britain (data from Robinson, 2009).

The traditional suppliers of the British appetite for bananas were the country's former colonies in the Caribbean, where conditions of employment and wage levels were relatively high compared to the 'banana republics' of Central America, where the US multinationals dominated the market. These countries were allowed to supply into markets at prices higher than those set by world commodity markets, under a special exemption from global free-trade laws known as the Lomé Convention, which was agreed between the then European Community and a group of producer nations in 1975. In spite of this arrangement, only 7 per cent of Europe's bananas originated from the Caribbean; US fruit corporations still controlled 75 per cent of the EU market. However, these companies were unhappy with the situation, and lobbied their government to challenge the Lomé Convention.

In 2001, the Clinton administration took a case against the EU to the WTO in a series of legal actions that became known as 'the Banana Wars'. Following prolonged wrangling, in December 2009 the EU agreed to cut the tariff it charges on banana imports, thus freeing access to European markets for US distributors. It is worth noting in connection with Clinton's speedy action that the legal challenge was launched within 24 hours of Chiquita making a US$500,000 donation to the Democratic Party. (A chronology of the events of the Banana Wars can be found at the BBC website: http://news.bbc.co.uk/1/hi/business/8391752.stm, accessed 13 September 2010.)

The Lomé Convention was intended to protect the conditions and wage rates of vulnerable workers in the poor countries of the Caribbean. To some extent, a similar end is achieved through the fair trade movement. Figures from the Fairtrade Labelling Organizations International indicate that consumers worldwide spent £1.1 billion on certified products in 2006 – an increase of 42 per cent on the previous year. Bananas were one of the crops that saw a particularly large increase, up 31 per cent. Another response has been for countries that produce bananas – amongst other commodities – to work together to increase their power in global markets (Lines, 2008).

Part of the purpose of this case study is to indicate that globalization is a political process and that, as a result of corporation lobbying of the WTO, trade patterns respond to profit-making rather than taking any account of local consumption patterns or the interests of workers or consumers. The question that arises with especial prominence in the age of climate change is how long Europeans can continue to favour a type of fruit that cannot grow in their climate and must be transported at the cost of a great deal of CO_2 emissions.

Summary questions

- What are the key features of globalization?
- Do you think that the long supply chains of the globalized economy are compatible with taking the action necessary to avert climate change?
- What is meant by 'the Banana Wars' and who won them?

Discussion questions

- Which aspect of the change brought by globalization do you value in your own life?
- Why do you think that it is easy for money to move across the world but not so easy for people to move to other countries to find work?
- Can you imagine what life would be like if 90 per cent of your resources came from your own bioregion?

Further reading

Friedman, T. (2005), *The World is Flat: A Brief History of the Twenty-first Century* (New York: Farrar, Straus and Giroux).

Hines, C. (2000), *Localisation: A Global Manifesto* (London: Earthscan): a polemical introduction to the environmental critique of globalization.

Mellor, M. (2010), *The Future of Money: From Financial Crisis to Public Resource* (London: Pluto): offers an explanation for the 2008 financial crisis as part of a view of globalization as a primarily financial phenomenon.

Stiglitz, J. (2002), *Globalisation and Its Discontents* (Harmondsworth: Penguin): an equally critical account of the uneven benefits of globalization and international development from a mainstream economist.

13 ⬤ Climate change: the greatest example of market failure?

13.1. The most serious issue of our time

It is the nature of science to be an area of contention and disgreement. Through challenging existing hypotheses and paradigms, young scientists make careers. They are, by nature, critical and contentious. It is striking, therefore, how little disagreement there is over the question of whether climate change is really happening and whether human activities are the cause. While a few scientific heretics maintain a range of bizarre theories in opposition to the mainstream view, the overwhelming weight of scientific evidence and opinion is that climate change is real, it is with us now, and that the way we behave as a species has caused it and is exacerbating it. The only reason that the sceptics are given so much attention is that it gives support to those who find the scientific conclusions so shocking that their only resort is denial.

Climate change is, quite simply, the most serious issue of our time. From the perspective of the human species, we might say that it is the most serious issue of all time because, if we do not take the necessary action to address it, we may not have a future as a species. This is a problem that is being caused by the economic habits of the industrialized nations whose emissions vastly outweigh those of the world's other countries. Table 13.1 gives data for emissions of greenhouse gases by the leading polluters in terms of per capita emissions. Figure 13.1 indicates the rapid and significant divergence in the levels of CO_2 in the atmosphere if we follow a business-as-usual path, compared with one that makes significant and rapid attempts to cut CO_2 emissions to stay within the 2°C warming scenario.

Although the overwhelming majority of the world's scientists are convinced that anthropogenic (i.e. man-made) climate change is a reality, publics in various countries are responding to this unpalatable message

What does the way forward look like? Two options among many: Business as usual or aggressive mitigation

Projected annual total global emissions (GtCO₂e)

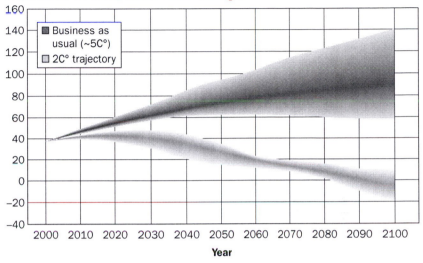

Figure 13.1 *Business as usual is not an option*

Source: World Development Report (2010), Figure 4. Thanks to the International Bank for Reconstruction and Development, for permission to reproduce this graphic, based on data from the World Bank

with the psychological response of denial – and as the evidence accumulates as to the seriousness of the problem, the level of denial is growing. A poll published in the UK *Times* on 14 November 2009,[12] just a month before the Copenhagen negotiations, reported that only 41 per cent of British people agreed that climate change is an established scientific fact; 32 per cent believed that the link was unproven; 8 per cent believed that warming was real but it was not caused by human activity; and 15 per cent believed that the earth is not warming at all.

Climate change is a problem that, although it became a central concern for environmentalists in the 1970s, has reached the economics community well into the twenty-first century. Section 13.2 details the findings of an economist who was asked to provide a policy framework for tackling climate change, and how this was greeted by his peers. The favoured response amongst economists was to attempt to create a market for carbon in the global market system – what they mean by this rather abstract idea is explained in Section 13.3. Section 13.4 gives space for the critics of market solutions to climate change, while Section 13.5

Table 13.1 The leading polluters: global roll-call of shame

Country	Metric tons CO_2e Per Person	Rank	$MtCO_2e$	Rank	% of World Total
Qatar [1]	66.6	1	58.9	73	0.16
United Arab Emirates	38.2	2	156.9	37	0.41
Australia	27.4	6	559	15	1.48
United States of America	23.5	9	6,931.40	2	18.33
Canada	22.9	10	739.3	9	1.96
Russian Federation	13.6	22	1,947.40	4	5.15
Germany	11.8	28	975.2	8	2.58
United Kingdom	10.7	38	645.3	10	1.71
Japan	10.6	39	1,356.20	6	3.59
China	5.5	82	7,234.30	1	19.13
Brazil	5.4	85	1,011.90	7	2.68
Nigeria	2.1	135	297.3	26	0.79

Note: Data are for total greenhouse gas emissions, reported as equivalent to the impact of CO_2. Gases included are: CO_2, CH_4, N2O, PFCs, HFCs, SF_6

Source: World Resources Institute, Washington DC: http://cait.wri.org

compares two policy proposals that seek to achieve fundamental structural change to the global economy as a necessary part of dealing with climate change.

13.2. The Stern Review: an economist encounters the environment

To those with a sceptical view of the relationship between the economy and the environment, it comes as no surprise that, once the UK government accepted the seriousness of the problem of climate change and required policy to address it, it turned to an economist rather than an environmental scientist. Sir Nicholas Stern (for he it was) was quite happy to admit his recent acquaintance with the problem of climate change when he was called upon to produce the most important international report into the issue in 2005. Unless we are to assume that Sir Nicholas was chosen on the basis that his name would convey the appropriate message about the seriousness of the subject of study, I think we might be forgiven for thinking that the government's mind was already made up that it was going to define climate change as an economic rather than a political, social, cultural or even spiritual problem.

However, the review panel sifted through the evidence they received from scientists and reached a conclusion that has not been popular with economists: the market system is failing and climate change is evidence of this. In discussion of climate change, the phrase 'business as usual' is frequently used (and abbreviated to BAU). This means continuing along the same economic path, with economic growth bought at the cost of greater use of fossil fuels, a scenario that Stern judged to be unacceptable. BAU would result in a certain increase in global tempreatures of 2°C by the end of this century and there is a 50 per cent chance that the increase would be by as much as 5°. Major change in the organization of our economic life is thus unavoidable, and the sooner that change occurs the cheaper it will be to achieve it. This is because the damage that climate change will bring with it will cost an increasing amount to rectify (we can think of damage caused to infrastructure such as roads and power lines as a result of more powerful storms or floods, for example). Stern predicts that this could cost as much as 5 per cent of GDP if you include only market impacts, and as much as 11 per cent if you include the negative impacts on health and the environment, which will also have to be repaired or healed. The conclusion of the Review is that the sooner we start reducing CO_2 emissions the less of this costly damage we will experience. Box 13.1 summarizes the most important findings of the Stern Review.

Box 13.1

Major conclusions of the Stern Review

- CO_2 emissions are caused by economic growth, but policy to tackle climate change is not incompatible with economic growth.

- Favours the transition to a 'low-carbon economy', which will 'bring challenges to competitiveness but also opportunities for growth'.

- 'Policy to reduce emissions should be based on three essential elements: carbon pricing, technology policy, and removal of barriers to behavioural change'.

- Argues for the pricing of carbon through trading, taxation or regulation.

- Need for government support for low-carbon and energy-efficient technologies.

- What we do now can have only a limited effect on the climate over the next 40–50 years; what we do in the next 10–20 years can have a profound effect on the climate in the second half of this century and in the next.

- By investing 1 per cent of GDP now (the next 10–20 years) we will avoid losing 20 per cent of GDP later (40–50 years).

- Markets for low-carbon energy products are likely to be worth at least US$500 billion per year by 2050, and perhaps much more. Individual companies and countries should position themselves to take advantage of these opportunities.

When Nicholas Stern described climate change as 'the greatest market failure of all time', he was being true to his disciplinary inheritance in suggesting that, if the globalized production and distribution market had functioned efficiently, then the problem would not have arisen. But what sort of market failure did he have in mind? One source of market failure is the problem of public goods, where the market price of a good that has been produced does not include the social benefits or costs that arise from its production. We could see climate change fitting into this definition, since we all share the benefits of the environment but do not pay for them. Since the use of the global atmosphere is not costed, this 'free good' is over-used as a result of the emissions of CO_2 in production and transport. Dealing with the public-goods aspect of the problem may be possible through the sorts of international negotiations that took place in Copenhagen in December 2009.

We saw in Chapter 3 that neoclassical economics describes the negative consequences of economic activity that occur outside the production unit as **externalities,** and this is one way we might conceive of climate change. Considering the climate-change consequences of CO_2 emissions as an **externality** helps to illustrate the problem neoclassical economics has with positioning the economy in relation to the environment, since while climate change might be external for the purposes of any particular factory, we are all observing more signs every year that it is very much an internal problem in terms of our environment. Policies may be developed to 'internalize' the externality by making it expensive to produce CO_2, and thus including its production in the cost curve of the firm. This would be achieved by creating a price for carbon: 'Carbon prices must be raised to transmit the social costs of greenhouse gases to the everyday decisions of billions of firms and people' (Nordhaus, 2007: 689).

The most contentious conclusion of the Stern Review – at least as far as neoclassical economists were concerned – was the choice of discount rate. After publication, the report was immediately challenged for suggesting that the costs would increase very rapidly as time went by, making it essential to introduce pro-climate policies as rapidly as possible. In making this estimate, the members of the Review team had to base it on some assumption about how costs would change over time, by using the standard orthodox procedure of introducing a 'discount rate'. As we saw in Section 3.4, working out the costs and benefits of any policy depends on the discount rate that is applied: the higher the discount rate, the lower the future cost of actions taken today. The Stern Review's conclusion that we need to act rapidly to tackle climate change resulted from his setting a very low discount rate.

Conventional economists were shocked by the consequences for the economy, and challenged this level of discount rate on the basis that it had exaggerated the effects of climate change in the distant future. Stern was basing all his conclusions on statistical models about the probability of events occurring. The possibility that the planet might cease to exist would clearly have a major impact on people's 'time preference', i.e. their preference for consuming now rather than in a (possibly non-existent) tomorrow. As Ackerman explains:

> Stern observed that a natural or man-made disaster could destroy the human race; he arbitrarily assumed the probability of such a disaster to be 0.1 percent per year, and set pure time preference at that rate. That is, Stern assumed that we are only 99.9 percent sure that humanity will still be here next year, so we should consider the well-being of people next year to be 99.9 percent as important as people today.
>
> (Ackerman, 2009: 86)

Nordhaus, a neoclassical economist, took issue with Stern's choice of this low level of discount rate, on the basis that it is unrealistic and it underestimates the ability of the economy to become more productive and solve the climate-change problem through technological advance:

> The logic of the climate-policy ramp is straightforward. In a world where capital is productive, the highest-return investments today are primarily in tangible, technological, and **human capital**, including research and development on low-carbon technologies. In the coming decades, damages are predicted to rise relative to output. As that occurs, it becomes efficient to shift investments toward more intensive emissions reductions. The exact mix and timing of emissions reductions depend upon details of costs, damages, and the

extent to which climate change and damages are nonlinear and irreversible.

(Nordhaus, 2007: 687)

Others have argued, to the contrary, that this quantitative analysis underestimates altogether the seriousness of the problem (Spash, 2007). It is also noteworthy that the conventional economists have focused so much of their discussion around concern for the appropriate discount rate rather than considering how the essential structure of the economy and increasing levels of consumption might be a more significant source for concern.

13.3. Pricing carbon: theory and consequent policies

Climate change is a difficult area for policy-makers for a number of reasons:

1. *Uncertainty*: There is a high degree of uncertainty about the problem itself (how much temperature will rise, over what time period, and what the consequences will be) and over the likely effectiveness of solutions in an area where there is no experience to base policy on.
2. *Credibility*: Policy-makers lack credibility, since citizens may well consider that their introduction of taxes, for example, is an attempt to raise revenue rather than to control climate change.
3. *Impracticality*: Effective policies to tackle climate change, such as the introduction of a scheme of personal emissions for the production of CO_2, are likely to be labour-intensive and thus costly on the public purse.
4. *Impersonality*: Whatever we do now, climate is likely to cause a deterioration in the situation we face for the rest of our lifetimes, which undermines our incentive to take action in our own self-interest.

Orthodox economists have come up with a range of market-based solutions to the problem of climate change, which focus on creating a price for carbon so that pollution is no longer a free good. Creating a carbon price will be a way of implementing the **polluter-pays principle** in the area of climate change – companies will no longer be able to treat the global atmosphere as a free dumping ground. Introducing such a cost would also create an incentive for polluting companies to invest in technologies that reduce their energy use and to switch to renewable forms of energy generation. So while there is widespread agreement with the Stern conclusion that 'Creating a transparent and comparable carbon price signal around the world is an urgent challenge for international collective action' (Stern, 2007: 530), there is considerable debate about the best way of creating that price, both in terms of efficiency and equity.

The first debate is about where the policy is implemented – this is the so-called upstream vs. downstream debate. Upstream we have the producers, so we might impose a tax on them, for example, when they extract fossil fuels from the ground. At the other end of the chain – downstream – we have consumers, whose emissions might be controlled through giving them a limited allowance per year, for example. Upstream solutions tend to be cheaper, since there are fewer producers, but how can we be sure that the costs will all be passed on to consumers? On the other hand, downstream solutions, involving millions of consumers, are expensive to administer but place the responsibility on citizens to change their individual behaviour.

Beyond that, the decision that needs to be made is fundamentally between a regulatory system, limiting and taxing CO_2 emissions, and a market solution that also imposes a limit, but then permits those who produce emissions to trade between themselves the right to do so. Those who support a system of emissions trading argue that it is efficient, since it ensures that those who make the reductions will be those who can do so most cheaply, and they will then sell their emissions rights to others, for whom that is a cheaper solution than reducing their own emissions. Such a scheme would (its supporters argue) also have the advantage that it would follow naturally from fixed caps negotiated internationally, and would provide a simple mechanism for governments to implement these caps nationally within fixed aggregate limits. However, following the failure of the Copenhagen negotiations, few would be sanguine about the likelihood of such binding international agreements being reached.

Clearly, such a case is most popular amongst more mainstream economists and businesspeople. The latter will see an economic advantage, since the rights to produce CO_2 that they will receive will constitute the creation of something of real value, and which they can sell, potentially increasing their profits. From a market perspective, trading will also be naturally self-balancing and will adjust in response to external price shocks, whereas taxes would remain fixed whatever might happen to, for example, the price of oil.

The other main policy proposal for pricing carbon is to introduce a form of carbon taxation. While carbon trading has gained more media attention and rhetorical support, the initial and most obvious policy to reduce CO_2

emissions is to tax them. The most popular proposal is for a tax that is applied as a fuel tax, based on the amount of fuel sold. When the fossil fuel is burnt, CO_2 is released and the quantity is directly related to the amount of fossil fuel consumed. The tax could be imposed in a number of different ways. The simplest would be an upstream tax, imposed on oil and coal companies when they extract the fuel from the ground. This would ensure that the total quantity of fuel is taxed and would be simple and cheap to administer. It would then be the responsibility of the fuel companies to pass the cost on to intermediate producers, who would then in turn pass the cost on to consumers.

The immediate appeal of a system of taxation is that it would address all polluters, not just the businesses who would become part of a carbon trading system. Although taxation systems are costly to establish and to monitor, they do not involve the transaction and negotiation costs that are present with any trading system. The advantage of a market system is that it would be self-adjusting, i.e. the price of a CO_2 permit would rise or fall according to demand. However, this could also be a significant disadvantage for businesses, because they would not be able to have a fixed idea about the cost of their emissions when producing business plans. There might also be a high degree of volatility in the price of CO_2 emissions, which could make planning difficult. A taxation system, by contrast, would be clear; it might be fixed on a gently rising trend so that businesses could plan for the cost of fossil fuels to rise gradually over time, and they could factor this in to their planning. Although such a cost would be unwelcome, it would at least be foreseen.

Perhaps the most attractive aspect of a taxation proposal is that it is a type of policy which is already familiar to both taxpayers and policy-makers. Creating carbon markets, by contrast, is an innovative and highly complex process. As is clear from the first experiment with such a policy – the European Emissions Trading Scheme (ETS) described in Box 13.2 – inexperience can lead to unexpected outcomes that may work against the objective of the policy. A tax would also generate revenues that could be reinvested in the infrastructure of a low-carbon economy – being made available as grants for home insulation or transition grants for businesses to install renewable energy systems, for example. This apparent 'benefit' is something of a double-edged sword, however, since the public is sceptical about pro-environment taxes, which they suspect may be introduced primarily to generate revenue rather than to tackle the environmental problem.

Box 13.2

The EU Emissions Trading Scheme

The EU Emissions Trading Scheme (ETS) was a bold attempt to apply neoclassical methods to the most serious market failure of all: climate change. The scheme involved issuing a number of permits to emit carbon dioxide, and giving them to 5,000 of the EU's biggest emitters, within a framework of the limits set by the **Kyoto Protocol**. The corporations that received the permits could then trade with each other, so that those who could more cheaply reduce their emission could sell the permits to those who found it more expensive to reduce theirs. It was estimated that the value of permits in the first round of trading was €170 billion: this shows that a huge value can be created when the global atmosphere is rationed in this way, and critics of the scheme have suggested that this value should have been widely shared, not allocated to a narrow range of corporations. In addition, firms have increased prices to reflect the pricing of CO_2 emissions, although they were themselves given the right to produce the gas free of charge. The World Wildlife Fund estimated that German utility companies will make windfall profits of between €31 and €64 billion from the scheme by 2012. The scheme was also criticized because only 43 per cent of EU emissions were included.

Any carbon-trading scheme is designed and implemented by politicians, and is therefore open to political influence at the national level – with Finland, Lithuania, Luxembourg and Slovakia all allocated 25 per cent more permits than their recent emissions would require – and at the local level, with powerful companies exerting influence on their governments to receive an unfair share of permits. Such systems are also rooted in the culture of business, and corporations have played a major role in designing the ETS; for this reason, it reflects corporate interests and only mildly constrains their activity. Perhaps most seriously of all, the ETS can encourage companies to keep polluting plants open – since, if they do not, they will lose their share of permits. The weakness of the original version of the ETS became clear in 2006, when the Scheme was on the verge of collapse because governments had given away so many licences that no company was required to do more than it would have done if the scheme had not existed. The price for the permits fell through the floor, incapacitating the market.

13.4. Taking the problem in the round

For critics, the idea of market trading to reduce CO_2 emissions is a symptom of the economic ideology fashionable in the twentieth century. This ideology suggested that markets were efficient problem-solving mechanisms, and that government control and intervention were to be eschewed. This explains the response to a pollution problem that is found in the realm of trading rather than legislation, as was the case with the 1956 Clean Air Act, which addressed a previous serious issue of air pollution in the UK. Critics raise serious questions about the usefulness, fairness and practicality of a market solution to the problem of climate change.

How can we be sure that the market analogy will extend to a virtual good like the global atmosphere?

Rather than proposing some form of trading as a solution to climate change, we might very well argue that climate change is evidence not of market failure, but rather of the weakness of the market as a basic distribution mechanism within the global economy. Rather than taking the market analogy into the field of climate change, perhaps we would do better to raise fundamental and critical questions about how the market economy functions and whether it is, in fact, the problem rather than the solution (Spash, 2010). The price system is a basic structure without which the market cannot function, but climate change means that the whole price system is in error:

> The problem lies with the whole economic process of business enterprise not some simple bilateral pollution problem which is a minor aberration of an otherwise perfect market system. Every product in the market place has embodied energy, is related to GHG emissions, and therefore has the 'wrong' price.
>
> (Spash, 2007: 709)

Even if we are convinced that market solutions have something to offer, we have only very limited experience of creating pseudo-markets for goods – like the right to produce CO_2 – that do not really exist. Our experience of creating pseudo-markets in the area of public services such as healthcare, transport and electricity supply have been mixed at best. Since our very survival depends on finding the right policy, we are taking a big gamble in assuming that we can create a functional market for CO_2 emissions.

Who establishes the market and sets the price?

As we saw above, the market for carbon is not like the market for carrots, where a large number of potential producers can find a patch of land and some seed and begin to grow their produce. In this case the 'product' – the right to global atmosphere – is poorly defined and politically determined. The permits to produce CO_2 that were sold in the EU ETS were created artificially by a power-bloc of Western states; the permits' existence relies on a system of policing and control that is not guaranteed. If a company exceeds the level of CO_2 for which it has 'bought' these permits, who will know? The market system's claim to superiority relies on the neutral system of price-setting, and yet in a carbon market the price would rise as a result of political decisions about the supply of permits. Such a process would inevitably be subject to massive political pressure, undermining any claim to scientific neutrality.

Are we all equally powerful consumers?

A system of trading carbon permits would be fine in theory, so long as the cap on the total quantity of emissions permitted was a strict one and all those trading had equal power within the market. However, this is clearly not the case. The first implementation problem of such thinking is deciding how the permits will be allocated. As described in Box 13.2, the ETS proved that any such system would be subject to massive lobbying pressure from corporate interests.

Any such system must be set within a global framework for CO_2 reductions, but the experience of Copenhagen makes it clear how difficult it will be to put such a framework in place. Negotiations foundered because the more industrialized and richer countries, whose citizens enjoy a higher standard of living, and produce more CO_2 as a consequence, argue that the share of emissions should be based on historical emissions levels (sometimes referred to as 'grandfathering'). Others, such as the Global Commons Institute in London, argue that the total emissions that can be produced should be shared fairly amongst the world's citizens on a per capita basis (how this would look in practice is illustrated in Figure 14.2 in the following chapter). This, they argue, is the only just allocation, and the only one that is likely to result in an agreement. Figure 13.2 indicates how unfairly CO_2 emissions are shared currently. Any global trading system based on equal per capita emissions would result in huge transfers of value from richer countries (which overproduce CO_2) to poorer ones. (Many have argued that energy-efficient technologies would be a good way of making this transfer.)

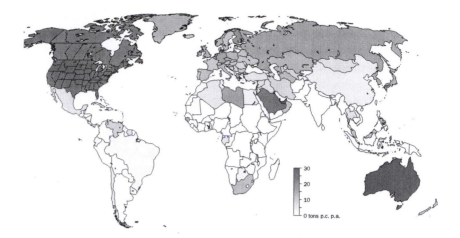

Figure 13.2 *Carbon emissions per person on a global basis*

Source: Reproduced freely thanks to Wikimedia commons

An additional problem with measuring emissions results from the fact that we need to include within our limits those emissions that were created to produce consumer goods – so-called 'embedded emissions'. Should these be included in the totals of the countries where the goods are produced, or where they are consumed? Figure 13.3 indicates the extent of 'indirect' emissions, i.e. emissions embedded in export goods, and those produced when they are transported across the globe. It does not seem right for China to be held responsible for emissions created when they produce TV sets that will be watched by American or European citizens.

13.5. Changing the climate or changing our lifestyle?

The solutions that have been covered so far are at the level of nations or large corporations. After a global agreement is reached, how should each country's legitimate emissions be shared between the people living in that country? If costs are imposed on producers and only affect consumers via prices increases, consumers with less disposable income will see a massive reduction in their standard of living. Two competing policies are being discussed to address this problem by allocating the right to produce CO_2 equally between citizens of a country. A **tradable emissions quota (TEQs)** operates like carbon rations, so you would need to spend some of your rations as well as money if you were to buy anything that had carbon embedded in it. The other system, called **Cap-and-Share**, would allocate

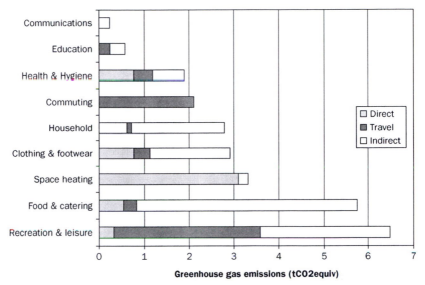

Figure 13.3 *The CO₂ emissions in all that we consume*

Source: Druckman, A. and T. Jackson (2010), Personal communication, 'The carbon footprint of UK households: Additional information', April 2010; Druckman, A. and T. Jackson (2009), 'Mapping our carbon responsibilities: more key results from the Surrey Environmental Lifestyle MApping (SELMA) framework', RESOLVE Working Paper 02-09, University of Surrey, Guildford, UK. Available from http://www.surrey.ac.uk/resolve/Docs/WorkingPapers/RESOLVE_WP_02-09.pdf

a permit representing the right to produce a share of CO_2 to each person. S/he could then decide whether to use it in burning up fossil fuel, sell it to somebody else, or destroy it. The designs, benefits and disadvantages of the two schemes are presented in Table 13.2. Both schemes begin by defining the global atmosphere itself as a 'commons', that is to say a shared resource that cannot be owned by private individuals (see more on this in the following chapter). They therefore consider that each citizen on the planet has an equal ownership right in the atmosphere and, by extension, an equal right to pollute.

When comparing these sorts of schemes, we can conclude that a personal allowance scheme forces each consumer to think very carefully about how they spend their ration of CO_2, and thus brings about responsibility and education. However, such a scheme is hugely complex and difficult to administer. By contrast, the Cap-and-Share system is easier to set up; however, it might be too abstract for people to grasp what it means to be given a carbon licence as an individual, and they might not be able to understand its importance. Any scheme like this, which creates an economic value through introducing a pseudo-market, can benefit those

Table 13.2 Carbon quota or Cap-and-Share?

	Tradable energy quotas	Cap-and-Share
Basis of sharing	Equal per capita shares	Equal per capita shares
Where is the cap enforced?	Downstream: Individuals and companies would need to surrender TEQ units in order to purchase fossil energy.	Upstream: Only companies importing or producing fossil fuels in the economy concerned would need to have permits.
Main advantages	1. The guarantee that the budgeted energy descent will be achieved.	1. C&S guarantees that any level of GHG emissions can be achieved by acting at the point at which fossil energy enters the economy.
	2. The assured ration of energy for individuals at a time of scarcity.	2. It shares the ownership of the atmospheric commons equitably and thus ensures that the burden of climate policy is also equitable.
	3. The long-term budget, which gives time to plan ahead.	3. The poor are compensated for both the rise in their personal fuel purchases and for the rise in the cost of energy.
	4. Specified in terms of energy (not money), so it involves everyone in energy-planning.	4. A long-term plan for tightening the cap gives a chance to plan ahead.
	5. Since TEQs (not money) are the *numéraire*, the system is resilient to the deep economic changes which are in prospect.	5. C&S works through the price mechanism and does not have a dual accounting system.
	6. The system depends on local and individual ingenuity to develop solutions.	6. C&S could build up into a system that provided a framework for a global climate treaty based on per capita shares and transfers of value from over-polluting nations to the poorer nations.
	7. It generates a common purpose between all participants – individuals, industry, the government.	

Source: The table is summarized from a briefing note prepared by the late Will Howard of Cap-and-Share based on information provided by FEASTA and David Fleming, originators of the two schemes

who are frugal in their use of fossil fuels, since they can sell what they do not use to others, and thus generate an income for themselves. Both schemes have the advantage of being fair, and also of creating a pressure to change lifestyles directly, rather than relying on the indirect mechanism of the price system. Hence, people will learn about how their consumption decisions relate to climate change, rather than just finding prices constantly rising.

Beyond this sort of discussion, we need to consider what it is about the way we live that is generating this vast amount of CO_2. Do we really need to consume in the way that we do, and is it actually making us any happier? This is a whole discussion in itself, and revolves around the issue of economic growth and whether it is a suitable aim for an economic system. Those arguments were presented in Chapter 9, and it is important to link that discussion to the solutions to climate change presented here.

Nothing could be more important than reducing our emissions of greenhouse gases. How we achieve this end depends fundamentally on the way our economy is structured, and whether the market system is part of the solution or the primary problem. The line being followed by Western governments was largely set by the Stern Review, which attempted to sell solutions to climate change as another growth opportunity: 'Tackling climate change is the pro-growth strategy for the longer term, and it can be done in a way that does not cap the aspirations for growth of rich or poor countries' (Stern, 2007: viii). Green and ecological economists, in contrast, are highly critical of this approach. They argue that changing our definition of prosperity and rethinking what a good life is for is a prerequisite to redesigning the global economy along sustainable lines. Such an economy would be radically different from the one we live with today.

Summary questions

- What were the main conclusions of the Stern Review?
- If climate change results from 'market failure', what is the nature of the market, and its failure?
- Which groups in society, or which types of country, will pay the highest costs of climate change?

Dicsussion questions

- How would you justify setting a zero discount rate for the damage caused by CO_2 emissions?
- If you were the CEO of an oil company, would you rather have a policy of carbon taxes or emissions quotas?
- If you were a pensioner on a low income, would you vote for a policy of Cap-and-Share or TEQs?

Further reading

Ackerman, F. (2009), *Can We Afford the Future? The Economics of a Warming World* (London: Zed): offers a very useful critical account of the neoclassical approach to climate change, with a particular focus on the key issue of discounting.

Nordhaus, W. D. (2007), 'A Review of *The Stern Review of the Economics of Climate Change*', *Journal of Economic Literature*, 45(3): 17: a view of Stern and his approach to discounting from the neoclassical side of the argument.

Spash, C. L. (2007), 'The Economics of Climate Change Impacts à la Stern: Novel and Nuanced or Rhetorically Restricted?', *Ecological Economics*, 63(4): 706–13: the author, a social ecological economist, gives his critical account of the Stern Review.

14 Markets or commons

Economics is concerned with the distribution of resources; for the majority of readers of this book, the dominant mechanism for distributing resources that they have known is the market. However, in other societies today, and during historical times in what are now the developed Western economies, different methods are used to meet needs, what we might refer to as 'provisioning'. This chapter will explore the idea of sharing resources through systems of common rights over land and its production, and will contrast this system with the market system we live with today, from the perspective of their impacts on the environment.

In a market system, allocation is derived from legal ownership. Property rights are clearly established and legally enforced, and these legal rights are considered to be the best guarantee of secure access to resources. For those who own little, or have difficulty in establishing their legal ownership, resources may be restricted, or insufficient to meet their basic needs. We might contrast this system with one based on need and use, as in the Marxist stricture, 'From each according to her ability, to each according to his need'. A similar principle was found in Roman Law: known as 'usufruct', it gave a person who does not own a piece of property the right to use it to meet his or her needs so long as the property remained undamaged. This principle has influenced approaches to land use in Latin America, for example the *ejido* system in Mexico (a system of communal land use), and has inspired the activity of land rights activists such as the MST (the landless workers' movement – see Section 14.4) in Brazil.

The modern discussion of the environmental impact of commons systems without clearly defined property rights was initiated by Garrett Hardin with his essay entitled 'The Tragedy of the Commons' (1968), which is discussed in Section 14.1. Section 14.2 then considers the historical system of commons as a means for sharing resources, and the destruction

common rights. Arguments were made to support this consolidation of land ownership in the hands of a few; these arguments revolved around the need for efficiency and to gain the maximum output from land. Some radical and green economists would argue that this emphasis on maximum productivity as the most efficient use of a resource is the origin of the ecological crisis that is being played out today. This process is continuing in the countries of the South, and the debate about the right to land ownership likewise continues across the world. The so-called Green Revolution has seen an increase in the productivity of land in the poorer countries of the world, but it has caused huge social upheaval and the destruction of social systems that relied on the pattern of common land ownership. So, although more food has been produced, this has not benefited the people whose hunger drove the change – the results of the new system of land use have been displacement, urbanization, unemployment, population increase and malnutrition (Shiva, 1991).

Having established the right to property in terms of physical resources, property rights have now moved into more abstract areas such as knowledge, through the development of intellectual property rights, and what are now known as 'global commons', such as the planet's atmosphere. In an otherwise critical account of his work, Fairlie credits Hardin with recognizing the ideological move of the privatization of common resources from the physical to the environmental:

> He recognised that the common ownership of land, and the history of its enclosure, provides a template for understanding the enclosure of other common resources, ranging from the atmosphere and the oceans to pollution sinks and intellectual property. The physical fences and hedges that staked out the private ownership of the fields of England, are shadowed by the metaphorical fences that now delineate more sophisticated forms of private property . . . Hardin must be credited for steering the environmental debate towards the crucial question of who owns the global resources that are, undeniably, 'a common treasury for all'.
>
> Fairlie (2009: 16–17)

The intellectual property regime has aimed to privatize some forms of life itself: generations of farmers and crop scientists have used selective breeding to produce species that are particularly suited to certain types of environment, or have qualities that are particularly valued. But global agribusinesses are being granted patents for these crops – some of which are basic to the livelihoods of poor people – in a process that critics refer to as 'biopiracy' (Hamilton, 2006). The neem tree is an example used by Vandana Shiva to illustrate how the process works and what she sees as its injustice (see Box 14.1).

Box 14.1

The neem tree as an example of biopiracy

The neem tree (*Azadirachta indica*) is a hardy, fast-growing evergreen that can reach 20 metres in height, and contains a number of chemical compounds that have been used by Indian people for generations. Its seed contains a chemical called azadirachtin, which is particularly valuable for its astringent properties. It is used medicinally to treat complaints including leprosy, diabetes and constipation; its twigs are used for cleaning teeth, and its oil is used to make toothpaste and soap. Neem oil is also a spermicide, and has been used as a contraceptive. Besides medical uses, it provides timber that is resistant to termites, and oil that can be used for oil lamps. After the oil has been pressed, the seed husks can be turned into feed cakes for animals. Neem also has insecticidal properties. This range of uses gained for the neem tree the Sanskrit name *Sarva Roga Nivarini*, or 'curer of all ailments'. It has also been cheap or free for most people to use, because they could extract what they needed themselves from an abundant local supply.

The extraordinary properties of the neem tree first came to the attention of Western scientists in the 1970s, and by 1985 the first patent had been registered and sold – to US multinational chemical corporation W. R. Grace. It saw a huge market opportunity in the production of natural pesticides. Since neem is a naturally occurring species, it cannot be patented, but pharmaceutical corporations either seek patents by synthesizing the active component, or patent the process they use to extract this component.

In a classic example of enclosure, Indian pharmaceutical corporation P. J. Margo, working with Grace, argued that the use of neem products outside the market was 'waste', whereas privatizing the production and processing would generate profit and hence wealth. In Shiva's own words, 'This statement is in turn a classic example of the assumption that local use of a product does not create wealth but waste; and that wealth is created only when corporations commercialise the resources used by local communities.' This exactly parallels the arguments made to justify the enclosure of English common land in the eighteenth century. The real outcome is that the price of seed has risen as profit-making companies have entered the arena, leaving it outside the range of the local people who developed its useful properties.

Source: Shiva (n.d)

These arguments about the impact of commodifying land and its natural processes, and turning them into products that can be traded (through their translation into what are defined as **ecosystem services** – see more in Section 10.3) is seen by some as a contemporary example of Enclosure. It is the corporations who are gaining from this process, since they have clearly established property rights, as well as the ability and finance to pay for legal services to defend and exploit these rights. Rio Tinto Zinc, for example, is exploring the profitable opportunities to sell ecosystem services that are available on the land they own. This includes 'potential biodiversity banks in Africa, as well as the opportunity to generate marketable carbon credits by restoring the soils and natural vegetation or by preventing emissions from deforestation or degradation' (Bishop, 2008: 11). In other words, the same companies that have created the CO_2 emissions are likely to profit from charging for their disposal.

This has been described as 'a major new wave of capture and enclosure of Nature by capital' (Sullivan, 2008b: 22). Nature is able to absorb the pollution from our productive activities, but only up to a certain point – and only if some land remains unexploited. That land belongs to the subsistence farmers and hunter-gatherers who have lived at a lower level of development and with a lower impact on the planet. Now these livelihoods are threatened, as those who have lived sustainably are displaced by the rush to exploit their land as part of a new commodity boom.

At the global level, the earth's very atmosphere is now being defined as a common resource – the 'global atmospheric commons'. As we saw in the last chapter, debate around the policy response to climate change is polarized between those who seek to commodify the atmosphere and create a pseudo-market to trade the right to pollute it, and those who see it as a commons that should be protected through socially determined and socially just agreements (Thornes and Samuel, 2007). If we treat the atmosphere as a commons, the right to pollute it must be shared equally between all the planet's citizens. Even if that right were to be commodified, the profits generated by the sale of permits to pollute would still need to be shared between the people of the world on a per capita basis (Barnes, 2001). This argument is behind the proposal for a Contraction and Convergence framework as the basis for an international agreement on climate change. As illustrated in Figure 14.2, this proposal suggests that, with a strictly enforced cap determined by the best available science, by 2030 countries should converge on an equal share per person of CO_2 emissions.

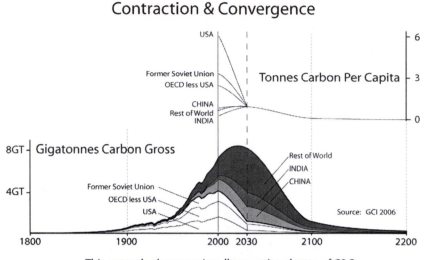

Contraction & Convergence

This example shows regionally negotiated rates of C&C.
This example is for a 450ppmv Contraction Budget, Converging by 2030.

Figure 14.2 *The Contraction & Convergence model for global CO$_2$ emissions reductions*

Source: Thanks to Aubrey Meyer and Tim Helweg-Larsen of the Global Commons Institute for producing and giving permission to reproduce this figure

So far this has been a very anthropocentric discussion. Some ecologists and bioregionalists suggest that, rather than focusing on reinforcing human property rights, we should consider the possibility of conferring legal rights on natural objects, such as stones or trees. They might be allocated guardians who would defend their interests during planning processes or when political decisions that might affect their future are debated (Chan, 1988). We might conceive of this as a 'parliament of all beings', extending our scope of moral concern and political rights beyond the human community.

14.4. Case study: MST, the Brazilian landless workers' movement

Practical challenge to patterns of private ownership has always been a live issue in Latin America, the most prominent example being the Movimento dos Trabalhadores Rurais Sem Terra (MST) or landless workers' movement in Brazil. Drawing on a history of peasant struggle for land during the 1950s and 1960s by groups such as the Ligas

Camponesas or 'peasant leagues', the MST began in October 1983, when a large group of landless peasants from across the state of Rio Grande do Sul in southern Brazil occupied a 9,200 hectare cattle ranch, which was owned by an absentee landlord. Over the following eight years, the movement staged 36 more occupations alongside protest rallies, marches and hunger strikes. They were supported by local radical priests and eventually succeeded in settling 1,250 families on their own land.

The MST is now the largest social movement in Latin America, with an estimated membership of 1.5 million people and a presence in 23 of Brazil's 27 states. This campaign took place in one of the most unequal societies in the world, an inequality exacerbated by the pattern of land use in which 1.6 per cent of the population are landowners and control nearly half of the nation's farmland, and 3 per cent of the population own two-thirds of the arable land. The MST organizes the occupation of unused land, which is then farmed cooperatively, with the construction of houses, schools and clinics. The campaign has been very successful, leading to the redistribution of nearly 30 million hectares of land; today some 45 per cent of Brazil's agrarian settlements are connected to the MST. The MST has achieved title to land for more than 350,000 families, and another 180,000 are waiting for the title to the land they have occupied. The campaign has used the principle of 'usufruct' enshrined in Brazil's constitution, which includes a clause that land that remains unproductive should be used for the 'greater social function'.

The MST has moved on from land ownership to challenge the neoliberal market model in other areas, with occupations of multinational corporation HQs, destruction of GM crops and political demonstrations. It has also moved along the supply chain, from primary production to processing and distribution, and has established about 400 business associations, including 49 agricultural co-operatives, 32 service co-operatives, 96 processing co-ops for goods including dairy, fruit and vegetables, cereals, coffee, meat and sweets, and three credit co-operatives. The campaign is not without risk – in April 1997, 19 workers were killed during a demonstration in Eldorado dos Carajás. For more on the MST, see Carter (2005).

Summary questions

- What was the 'tragedy' that Hardin foresaw for the medieval commons, and what present situation did it prefigure for him?
- What do you understand by the concept of a global atmospheric commons?
- In what sense can intellectual property rights be considered to be a form of 'enclosure'?

Discussion questions

- Do you think the issue of land ownership has anything to do with achieving environmental sustainability?
- Why might a landless peasant in Indonesia and the CEO of a global oil corporation view the issue of common resources differently?
- Will the people of India be able to use better toothpaste if the neem tree is patented or if it remains a common resource?

Further reading

Fairlie, S. and Fernandez, J. (eds) 7 (2009), 'Dismantling the Commons: A History of Enclosure in Britain', special issue of *The Land*, summer (Flax Drayton: TLIO): a non-academic introduction to various issues surrounding land ownership and enclosure.

Hardin, G. (1968), 'The Tragedy of the Commons', *Science*, 162: 1243–8: the seminal paper that introduced the concept of commons into environmental debates.

Ostrom, E. (1990), *Governing the Commons* (Cambridge: Cambridge University Press): a scholarly work on social systems that may offer an alternative to markets for the distribution of some resources.

15 Conclusion: Is it the economy? Are we stupid?

During his presidential election campaign in 1992, Bill Clinton famously had a sign with the phrase, 'It's the economy, stupid' hanging in his campaign headquarters. In 2008, some of us who are deeply concerned about the collision course human civilization is running with the planet also resorted to calling each other stupid, with the release of the film *The Age of Stupid*. I am in the camp that thinks it is not a very good idea to call each other names.[13] For one thing, it is rude and undermining. For another, the clash that has been developing between economy and environment is not because of stupidity. We would have to be very stupid indeed to risk the future of our lives, our descendants' lives and the beautiful planet we love, just for so many flashy gadgets and pieces of plastic crap. Yes, the way our economy works is the fundamental cause of the environmental crisis, but the fact that we are struggling to change our economy is not because of stupidity but because of an unequal distribution of power. Not to recognize this is to undermine our ability to make the changes that are necessary to safeguard human life on earth.

There is no question that we are facing a critical moment as a species. Lord Stern has called climate change 'the greatest market failure of all time', and it has been generally agreed that it is a more serious problem than terrorism. But climate change is just the most pressing of a range of issues that give evidence of the way our economic system is out of balance with our environment. Whether we think of habitat and species loss or the depletion or a range of resource, whether we consider the rise in pollution-related disease or the deterioration of the soils we rely on to grow food, across the world we have a mass of evidence that our planet is in peril as a result of our activity.

15.1. Too clever for our own good?

Humans are ingenious and cooperative, and it is these qualities that we need to exploit to the full now in redesigning economic systems so that they meet our needs without endangering our planet or the ability of our descendants to live comfortably as part of it. As a green economist, I do not feel particularly warmly towards technology – and I am wary of using it as a way of avoiding the structural change I believe is necessary – but this does not undermine my faith in the ability of clever, engineering minds to facilitate our journey towards a future that is environmentally sensitive. What we are seeking especially is cleverly designed processes and systems that achieve effective outcomes without the use of fossil fuels. Because we are boxed in by climate change, we need technology driven by wit rather than power.

An example of such a technology is a cement substitute called Novacem,[14] which solves a problem with our dominant construction technologies that has only recent been recognized. Not only are building materials such as bricks and cement highly energy intensive in terms of their manufacture, but in the case of cement, because of its chemical structure, it also produces carbon dioxide throughout its life. An early solution was the use of the more primitive lime render, which can absorb a high proportion of the carbon emissions that are generated when it is produced, and requires less fossil fuel in its production. But a more advanced design has led to the development of Novacem, which is made from magnesium silicates, and hence actually absorbs CO_2 during its working life as a building material. Novacem draws CO_2 out of the atmosphere while it hardens, and is thus the first of what may well be a whole generation of 'carbon negative' construction products.

Novacem is still in the early stage of development, and so has yet to prove itself in terms of strength, durability and the energy needed to manufacture it, but it would appear to be the sort of product that can actively combat climate-damaging emissions, rather than produce them. A whole range of such techniques and methods also needs to be developed in the agricultural sector, redesigning farming so that land can be used as efficiently as possible to sequester carbon, rather than farming relying heavily on fossil fuels and contributing to the problem. So technological and especially design solutions have an important role to play in easing the relationship between the human community and the planet, but they are not a substitute for structural change in the economic model that dominates twenty-first-century life.

It is generally assumed that efficiency of resource use will lead to a lower level of environmental impact, most commonly in the case of energy efficiency, which is suggested as a major policy response to the problem of climate change. However, this is an assumption that fails to take into account the human factor and the range of possible responses that people might make to higher efficiency levels. An example of an unexpected response is the 'rebound effect', where the response to increased energy efficiency is to use more of the good or service that is now produced more efficiently. For example, if a home is better insulated, the people who live there may choose to enjoy warmer surroundings, rather than use their central heating less, and thus some of the efficiency will not translate into energy savings. Box 15.1 reports evidence for the rebound effect resulting from the redesign of the domestic washing machine.

Box 15.1

Direct rebound effects for clothes washing

Research by Davis (2007) provides a unique example of an estimate of direct rebound effects for household clothes washing – which together with clothes drying account for around one-tenth of US household energy consumption. The estimate is based on a government-sponsored field trial of high-efficiency washing machines involving 98 participants. These machines use 48 per cent less energy per wash than standard machines, and 41 per cent less water. While participation in the trial was voluntary, both the utilization of existing machines and the associated consumption of energy and water were monitored for a period of two months prior to the installation of the new machine. This allowed household-specific variations in the way people use their washing machines to be controlled for, and allowed the researchers to estimate how much their use varied as the price of using the machine changed.

The monitoring allowed the marginal cost of clothes washing for each household to be estimated. This was then used as the primary independent variable in an equation for the demand for clean clothes in kg/day. Davis found that the demand for clean clothes increased by 5.6 per cent after receiving the new washers, largely as a result of increases in the weight of clothes washed per cycle rather than the number of cycles. While this could be used as an estimate of the direct rebound effect, it results in part from savings in water and detergent costs. If the estimate was based solely on the savings in energy costs, the estimated effect would be smaller. This suggests that only a small portion of the gains from energy-efficient washing machines will be offset by increased utilization.

> Davis estimates that time costs form 80–90 per cent of the total cost of washing clothes, and that the larger the quantity of clothes that can be washed at a time, the more consumers feel they are saving time. The results therefore support the theoretical prediction that, for time-intensive activities, even relatively large changes in energy efficiency may have little impact on demand. Similar conclusions should therefore apply to other time-intensive energy services that are both produced and consumed by households, including those provided by dishwashers, vacuum cleaners, televisions, power tools, computers and printers.

Figure 15.1 illustrates the intricate relationship between product design, improvements in design processes, economic structure and social responses that are all involved in productive activity. Reducing the energy required to produce any particular good will lead to a fall in its price, which may then stimulate further demand and increase the sales of that product. This system of growth in the productive economy can be conceived of as a reinforcing feedback cycle, with technological advance being one of the factors that drives economic growth (and hence resource and energy demand), rather than reducing demand and promoting conservation of energy and resources as in the standard neoclassical supply-and-demand model.

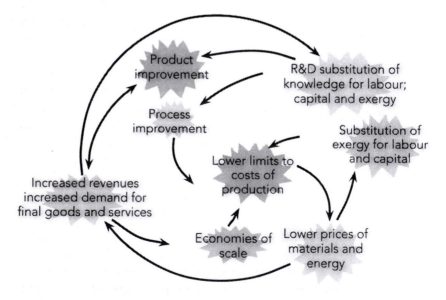

Figure 15.1 Complex feedback system of economic growth

Source: From Figure 6.1 in Herring and Sorrell (2009); redrawn by Imogen Shaw

Rebound effects point to the difficulty of devising policies to tackle environmental problems. They also indicate that the major changes that need to occur as we move towards a lower-carbon economy will be in terms of a social and cultural paradigm shift rather than a technological change. So having citizens who appreciate the depth of the environmental crisis and are engaged in changing their lifestyle to contribute to its solution is more important than paying scientists to devise gadgets or even to redesign production systems.

15.2. Putting the market in its place?

I have begun this concluding chapter with a discussion of the rebound effect because it is a neat exemplification of the two tensions running through this book and through the relationship between the economy and the environment. The first is the tension between the market and society as the appropriate setting for tackling the environmental crisis. The second is the tension between whether we need to change the direction of travel of the supertanker that is the global economy, or whether it needs a fundamental redesign. These two tensions enable me to summarize the contributions of the different economic approaches outlined in Part I.

It is, of course, risky to generalize about the life's work of hundreds of academic economists, but it is largely true to say that the contributions considered in Chapters 3–5, those of the neoclassical, environmental and (for the most part) ecological economists, accepted the structuring of economic life by the market. The role of the environmental economists, in particular, is to remain true to this belief in markets even in areas where it is apparent that they are failing spectacularly (such as on climate change) and to consistently propose more markets, even for non-existent goods such as 'the right to produce carbon dioxide'. By contrast, the economists whose work is presented in Chapters 6 and 7 identify the capitalist economy as *the* source of the environmental problem. It is not that there is an inevitable tension between *any* economy and the environment, but there can be no peace between a capitalist economy and the environment.

In spite of our politicians' recent affection for a consensus approach to life, this is a fundamental and unavoidable difference of opinion. You cannot have an expansionist globalized capitalist system whose central logic is growth for just part of the world, or for some goods, or just on Saturday mornings. If this is the system you choose, then you have to trust that same system to find solutions to the environmental crises. That is not to say that we have to abolish private property and move to a system of state control; far from it. As we saw in Chapters 6 and 7, the proposals from the economists who reject the market system are, in fact, based on a greater diversity of economic forms and a more participatory democracy than those proposed in Chapters 3 and 4.

The rebound effect helps to explain why this disagreement is so stark. Only a few of the hardest-line neoclassical economists still maintain that there is no problem with energy use, resource depletion and waste, but their solution to these problems is technological advance. To propagate this idea, they have come up with the concept of 'decoupling' – detaching economic growth from a greater use of resources and particularly energy. As I outlined in Section 15.1, I have no argument with this. If a clever young scientist could invent a perpetual motion machine that removed the problem of climate change in a puff of smoke, nobody would be happier than me, because I have children too, and I hope one day to have grandchildren. But the laws of physics – which, unlike the laws of the market, are immutable – tell me that this is impossible. Yet while the laws of the market may be subject to change, the central law of a capitalist economy – its need for perpetual growth – cannot be changed. This is why technological developments within a capitalist economy will always be used to generate more growth. It is a change in the design of the economy itself that we need, not in the design of any one of the thousand different products that the economy invents every week.

At the personal level, the rebound effect tells us that, within a society where our ethic of consumption is a reflection of the wider growth ethic and we have been trained to respond to price signals, our response to energy efficiency is likely to be to use products more. At this level, many of us are beginning to take responsibility and realize that the most important change we can make is not in the kind of lightbulb we choose, but rather in our whole wordview. This is a helpful step, but it is not sufficient for us to achieve a comfortable relationship with the planet – while the economic system itself is dominated by the growth dynamic we will always be swimming upstream. A market economy will use energy efficiency to generate more products more cheaply, and therefore increase the number of products available to be bought and sold. This is the rebound effect writ large, and its only solution is to completely restructure how our economy works, so that the growth imperative is replaced by an ethic of sustainability.

Herman Daly, a former economist at the World Bank who was one of the pioneers of ecological economics, compared the growth-based economy to an aeroplane, which has to keep moving forwards if it is to stay in the air. What we need to do, he argued, is to redesign it as a helicopter: the steady-state economy, which can remain airborne without forward motion, or economic growth. However, as the Canadian ecological economist Peter Victor has pointed out, the problem is that we need to convert the aeroplane into a helicopter while it is still in the air. This is the extent of our challenge: how to make significant structural changes to our economic system without catastrophically undermining its ability to provide for the 7 billion people who live on this planet. As a green economist, my response would be that this task would be considerably less challenging if we were not all sitting in the same flying machine, whatever its design. Our challenge is to build a diversity of alternative, locally grounded, self-reliant and resilient economies that rely on balance rather than growth, and can, if you will allow me to mix my metaphors, act as lifeboats for the passengers of the doomed and destructive capitalist vessel.

15.3. Whose common future?

The first official UN report to alert humankind to the dangers facing our planet was published by the Brundtland Commission in 1987. It was called 'Our Common Future'. In one sense, this was a good title: if we are to have a future on this planet then we need to think about sharing. But the title concealed as much as it revealed, because it suggests that we have equal power to determine our destiny, and that is far from true. According to Pepper:

> [W]e sometimes underemphasise the importance of material, economic, vested interests in shaping processes of global modernisation and global *ecological modernisation*. Yet since the wave of neo-liberalisation beginning in the 1970s, the central role of such interests in social, political, cultural and ecological spheres has been very evident for all to see.
>
> (Pepper, 2010: 42)

The economists whose work I have covered in this book differ in the extent to which they explicitly address the relationship between the economy and politics. For the neoclassical economists, the market is an objective system whose laws are immutable and superior to the grubby world of political actors. Environmental economists take a similar view; even the ecological economists, while more sceptical of existing economic structures, see the solutions in scientific rather than political arenas. It is the economists in the green and anti-capitalist schools who see the power structure as an inherent part of the economic problem, and it must have become clear by now that this is also my view. The recognition that the earth's resources are limited underlines the importance of sharing the remaining resources, and, unless we can find a way of negotiating this sharing, we face a future of conflict over resources – whether these be the diminishing oil supplies, the stores of minerals that are being consumed ever more rapidly, or even basic resources such as agricultural land and water. So it would appear that we cannot solve our problem of sustainability without also addressing the weakness of democracy – locally, nationally and globally.

The negotiations at Copenhagen have made clear – if it was not already clear enough following the demonstrations against the World Trade Organisation or the failure of the Doha negotiating round – that the countries of the global South are no longer prepared to accept a division of the spoils where the terms of trade work against them. The globalized economy was developed from colonial systems that assumed the power vested in the countries who had developed sophisticated technologies gave them a right to a greater share of the earth's resources. The closure of the planetary frontier makes the issue of sharing resources on a global scale more salient. The other side of the coin is the leadership shown on the issue of climate change by some of the larger and more successful countries of the South, particularly Brazil and China. The future of the human species on planet earth will inevitably be a common future, and this has major implications in terms of quality of life for those of us who have thus far had access to a larger-than-fair share of the earth's resources.

The discussion of climate change in Chapter 13 highlighted the disagreement about whether the market system is the potential solution to the environmental crisis or the real cause of the problem. I have made it clear throughout this book that I am sceptical about the role that markets can play. I hope that you will make up your own mind, and follow up on some of the opposing positions that have been presented throughout this book. Whether or not you choose to follow the theories of neoclassical economists, I think it would be hard to argue that this is the only approach to economics that has any merit. In the opening chapter, I discussed the development of economics as a discipline, giving a very brief sketch of how we have arrived at the current position in terms of how we analyse the environment.

Figure 15.2 indicates the attitude towards the environment of the different schools that can be identified today. It makes clear that each takes a quite different view of the nature of the planet and its resources. Far from a

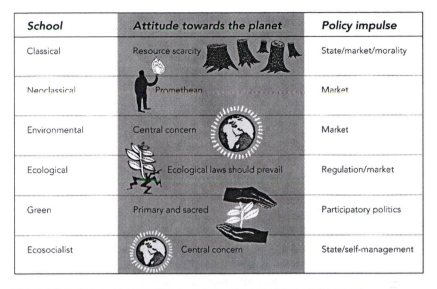

School	Attitude towards the planet	Policy impulse
Classical	Resource scarcity	State/market/morality
Neoclassical	Promethean	Market
Environmental	Central concern	Market
Ecological	Ecological laws should prevail	Regulation/market
Green	Primary and sacred	Participatory politics
Ecosocialist	Central concern	State/self-management

Figure 15.2 *Economic paradigms and the placing of the environment*

Source: Author's graphic drawn by Imogen Shaw

rational, scientific decision based on evidence, these views are personal and culturally specific judgements, which have changed radically in the past – and can be changed again. The first economists were deeply concerned about the scarcity of resources, and whether we would be able to feed the growing population. The advent of fossil fuels and rapid growth in scientific knowledge enhanced our ability to exploit resources, leading to an optimistic and almost hubristic approach to nature, which I have compared with that of the Greek mythical character Prometheus. This optimism waned in the second half of the twentieth century, as scientific evidence demonstrated the negative consequences of an approach to nature that was heedless of limits. The most radical economists covered in this book – the green and anti-capitalist economists, both of whom identify the economic system itself as the primary cause of ecological crisis – differ in the nature of their commitment to sustainability. The green economists are much more likely to identify the origin of their work in a spiritual commitment, which contrasts with the more materialist orientation of the anti-capitalists.

Figure 1.1 illustrates several 'lost traditions' of economics that have informed the work of economists who are concerned that the environment and economy are on a collision course. My own view is that we need to allow this heterodoxy into the debate about the future direction of economic policy, and to accept that new approaches to economics that retrieve insights from some of the work that is rarely considered today may have much to offer us. Most importantly, we should challenge the domination of both research and teaching by one narrow and very historically specific branch of the discipline, that of neoclassical economics.

15.4. An over-supply of bad news?

As discussed in Chapter 3, the dominant economic paradigm views people as rational economic beings, yet our response to the environmental crisis suggests that this may be a distorted view. Responses to climate change in particular demonstrate a range of behaviours that, while entirely psychologically normal, are far from rational. The science of the case is clear: the planet is warming and human activities are the cause. Yet all sorts of highly educated and generally well-informed people do not believe this. They are taking the normal human response to very bad news: denial. This is not unhelpful or unexpected – the ability to protect ourselves from shocking and sudden bad news in this way enables us to survive psychologically – but it is not rational, and it is not helping us find solutions to our problems. When I write about natural human responses like these, I should make it clear that policy-makers and economists are – for all their straight-line planning and conception of people as rational – just as susceptible to these responses as anybody else.

Another natural response to shocking news is displacement activity. So when faced with the prospect of danger, a wild animal might begin to preen, or scratch one part of its body obsessively. In the case of human beings facing climate change, we might decide it is terribly important to redecorate the spare bedroom or buy a flat-screen TV. In this case, it is particularly unhelpful that much of our displacement activity is encouraged by those who profit from it, and is powered by fossil fuels. In view of this human propensity to be least capable when the threats are greatest, those who can grasp the reality of the situation, understand the implications, and remain capable of acting in a constructive and positive way are especially valuable – our survival depends on them.

While I am on the subject of psychology, I should mention a third insight from that discipline, which has increasing relevance: cognitive dissonance. James Robertson drew my attention to the prevalence of this, and its creation by our political leaders. Cognitive dissonance is the psychological stress caused by being required to believe two mutually incompatible things at the same time – an example is perhaps how Stalin's mother felt when she realized that the little cherub she had nursed had become a mass murderer. More relevantly, we can consider the example of a government that pays for TV advertising admonishing us to drive five miles less per week, while at the same time giving permission for a third runway at Heathrow Airport.[15] These mixed messages – a result of governments finding themselves caught between corporate power and scientific irrefutability – add to our stress, thus reducing our ability to act rationally and constructively, and also reducing the credibility of the very people we are relying on to make the political changes that we need. Alongside the abnegation of political power in favour of the power of economic forces, the absence of leadership, in the face of a crisis they themselves draw attention to in emotionally threatening messages, is deeply unsatisfactory.

So my conclusion is that we should be realistic about the nature of the problem and the nature of the human beings who have to solve the problem. We are all affected by the environmental crisis in a deeply personal way, and we are emotional and intuitive *human* beings with the capacity to display extraordinary powers of rationality, as well as courage and resilience. We will need *all* these qualities in the coming century if we are to protect our planet and our species, and we should have the humility to accept our diversity as a strength rather than a weakness. When Darwin wrote of species survival, he used two phrases that have passed into our culture of understanding the world: 'survival of the fittest' and 'natural selection'. The first is often misinterpreted, and particularly so by economists: the word 'fit' in this context does not mean trim and in good health, it means appropriate (as in the now ubiquitous cliché 'fit for purpose'). It links to the second phrase, because the selection that takes place is from a wide-ranging diversity of forms, from which nature chooses that most suited to survive in the given niche. One of the most dangerous aspects of our contemporary culture – particularly in economics – is its concentration on uniformity, and especially the hegemonic domination of the market. Creative responses to crisis have always required us to use the variety at our disposal; in planning an economy that is not on a collision course with its environment, taking a heterodox approach would allow us to be responsive to the unexpected and unpredictable events that we will have to face.

We will need our policy-makers and politicians to abandon their projections and accept the humility of following rather than leading nature. We will need them to accept that the world is constantly emerging into new forms, and that, because it is in the nature of human beings to be cunning and to subvert their policies (see the example of rebound effects discussed in Box 15.1), they will need to be adaptive, compassionate and imaginative. These are the qualities – the cunning as well as the compassion – that have enabled our species to survive and thrive for thousands of years. Rather than an MBA or an ability to respond well in media interviews, what we require of our politicians and policy-makers is courage, humility and leadership.

15.4. It's your future; it's your choice

I have written this book primarily with a young readership in mind; I work in a university, and have been astonished by the poor level of knowledge about the effects of the economy on our environment that I have found amongst my students. In all the debate over what an educational system is for, nobody seems to have considered its role in ensuring that we have a viable and enriching experience as human beings sharing the planet to have been a priority. As well as the content of the curriculum, I find myself apologizing for the state of the planet that they find themselves in as they move out of education and into their role as citizens. My generation has left the world in a sorry state, and we have bequeathed to our children the task of curing the ills that our lifestyles have created.

As is made clear throughout this book, I have a very particular view about what the best remedies are, but it is up to you, the twenty-first-century citizens, and those who will still be here to experience life at the 2050 date that is just a mythical future for most of the Copenhagen negotiators. What sort of life do you want to see? We can imagine a whole range of terrifying dystopias where food shortages lead to famines, wars and even the recycling of human beings into soylent burgers,[16] and where rising temperatures result in starker divisions between the wealthy – inhabiting gated communities safely protected from severe weather events – and the poor, who are reduced to climate-induced migration. Or we can have faith in the technical experts to design systems and products that enable us to maintain a similar lifestyle standard, sharing it with the growing world population and powering it without the use of fossil fuels.

My own conviction is that the route to a positive future will rather be one that is shared between all the world's people. We will need democratic involvement and a diversity of skills and knowledge if we are to maintain our cultures and our humanity. The future that will become yours depends on the action you take now. I hope that after reading this book you have come to share my view that, in finding our way towards a happier, healthier and more balanced future, the role of the economist is crucial. At its heart, economics is about resources, and so, since all resources are ultimately derived from the environment, it should play an important mediating role between people and that environment. The fact that it has been reduced to the dismal science of projection and regression should not be allowed to deter those with more colourful concerns from entering the field. Economics is due for a revival, and a wealth of diverse cultures, views and ideals is needed to bring that revival about. Whichever type of economics appealed to you most while reading this book, I hope you will use it to join me in reframing economics as a creative and hopeful journey.

Summary questions

- What do you make of the rebound effect? Can you think of an example of how you might use more energy because of an increase in energy efficiency?
- Why do you think so many people still deny that climate change is caused by human activity when the overwhelming majority of scientists have concluded that it is?

Discussion questions

- Which of the approaches outlined in Part I did you find most constructive?
- Do you think that the environment–economy tension requires structural changes to the economy?
- If solving the ecological crisis is a priority for you, do you think it would support your mission to learn more about economics?

Further reading

Herring, H. and Sorrell, S. (2009), *Energy Efficiency and Sustainable Consumption: The Rebound Effect* (Basingstoke: Palgrave): discusses the question of the rebound effect as it relates to the environmental crisis.

 # Glossary

Words defined here are printed **in bold** in the text.

Bioregion A natural region defined by its ecological coherence, rather than by political boundaries.

Bio-prospecting Exploring the biosphere (the planet and its species) in the hope of finding plants or animal species that can be profited from.

Cap-and-Share (C & S) A policy to tackle climate change that issues permits to produce CO_2 (or the value of them) to citizens on an equal basis.

Capital, financial Financial assets that can be used to provide an income.

Capital, human The skills and abilities of individuals, gained through education or experience, which can be used to support economic activity.

Capital, manufactured Physical assets that can be used as the basis of profitable production.

Capital, natural Naturally occurring resources that can be exploited to make a profit.

Capital, social Social networks of trust that can support economic activity.

Carbon offsetting Investing in carbon-reducing activities to compensate for the production of CO_2 while still maintaining a high-energy lifestyle.

Carbon sinks Natural systems that extract CO_2 from the atmosphere and trap it, especially forests.

Debt-for-nature swaps Financial agreements where a rich country pays a poor one not to destroy its natural environment.

Discounting A process that enables future costs and benefits to be given prices that are equivalent to comparative present prices.

Ecosystem services A phrase used to describe all the unquantifiable benefits that nature provides free of charge.

Embodied energy The energy used in the production of a product.

Endogenous Arising from within a system, e.g. an excessive growth of lice may cause damage to a fish farm, but this is caused by the existence of the fish farm itself, as it is only living in such close proximity that fish will develop such an infestation.

Externality The impact of a process on a third party that is not included in the price paid by the consumer or the costs of the producer.

GDP (gross domestic product) A measure of the goods and services produced within the borders of a country, whoever owns the companies that produce them.

GNP (gross national product) A measure of the goods and services produced by companies owned by the citizens of a country, wherever in the world it is produced.

Green New Deal A plan to invest in the green economy to solve the problems of recession and climate change simultaneously.

Heterodox economics The dominance of the neoclassical paradigm within academic economics has led to all other approaches being labeled 'heterodox' as opposed to this orthodoxy.

Holistic Concerning itself with the whole system rather than breaking it down into parts.

Kyoto Protocol A UN convention negotiated in 1997, which included legally binding targets to cut CO_2 emissions.

Marginal The effect of a small change in any variable; effectively, the effect of the last unit of production, labour, etc.

Macroeconomics The large-scale consideration of a national economy, or interaction between economies.

Microeconomics Consideration of the economy at the small scale of consumer choice or the behaviour of firms.

Net (benefits/costs) An indication that something has been subtracted, usually a potential increased value due to interest accumulation or decreased value due to inflation or depreciation.

Non-renewable resources Resources that cannot be replaced once they are exhausted, especially fossil fuels.

Optimality (Pareto) A situation where the best outcome is achieved, especially in sharing goods or resources, so that nobody can be better off without making somebody else worse off.

Precautionary principle In a situation where there is uncertainty as to the impact of a process, the burden of proof lies with the person who would introduce the process, to prove that it is not harmful.

Polluter-pays principle The idea that the organization or nation responsible for pollution should pay for the clean-up.

Reductionism Analysis that involves breaking something down into its parts rather than considering the system as a whole.

Renewable resources Resources that are available on a continuing basis, e.g. tidal power.

Rent (technical term in economics) Income from allowing somebody else to use your land or property, and by extension (and pejoratively), the income that is acquired as pure surplus.

Sustainability/unsustainability The characteristic of systems that will not only endure, but will do so without diminishing the quality of the environment.

Tradable emissions quota (TEQ) A scheme to allocate a per capita share of CO_2 to each person per year, which they are free to choose how to spend.

Utility A technical way of considering welfare: a higher rate of satisfaction would be indicated by a higher value of utility.

 # Notes

1   World Development Indicators Database:
http://data.worldbank.org/indicator/IS.VEH.NVEH.P3 (accessed 13 September 2010).

2   Thanks to Dr Steve Harris of Glamorgan University, who introduced me to the work of Howard Odum and wrote this section.

3   Thanks to Ioana Negru for drawing my attention to this quotation.

4   This rigid distinction between a 'use' and a 'non-use' value illustrates how strongly the neoclassical paradigm, with its focus on the maximization of **utility** as the central aim of the economic agent, dominates environmental economics. A **heterodox** economist might question why eating a sandwich represents something 'useful', whereas enjoying a view does not.

5   This definition between weak and strong **sustainability** is an important one. Since sustainability is concerned with whether human needs will continue to be met from the earth's resource in the long term, the significance placed on **non-renewable resources** is of vital importance. According to proponents of weak sustainability, we need not concern ourselves with the strict limits on non-renewable resources, since we can create man-made substitutes. Those who subscribe to the strong sustainability view argue, to the contrary, that we must maintain the stock of **natural capital**, that is the value of the earth itself, because there are not substitutes for these natural systems that we depend on.

6   Thanks to Dr Steve Harris of Glamorgan University for co-writing this section.

7   **Social capital** is a measure of the levels of trust within a community or society, whereas **human capital** is an economic term to express the skill level of the people within a firm or society, which is frequently equated with educational qualifications.

8   In spite of his criticism, he stands for the US Green Party, and ran against Ralph Nader for the US Green Party presidential nomination.

9   The film *The Power of Community: How Cuba Survived Peak Oil* is an account of this transition: http://www.powerofcommunity.org/cm/index.php (accessed 13 September 2010).

10   **GNP** and **GDP** are two slightly different ways of measuring national economic production. The 'gross' means that both measures do not allow for the loss of value that occurs over time as equipment wears out. The 'national product' is production of companies registered in the country under consideration, wherever in the world they are located, whereas the 'domestic product' includes only what is made within the borders of that country, whether or not those companies belong to the citizens of that country.

11   A report of this poll can be found here: http://www.timesonline.co.uk/tol/news/
     environment/article6916648.ece (accessed 13 September 2010).
12   You may enjoy the Stand-Up Economist's interpretation of the principles of
     neoclassical economics, which includes several suggestions that people are stupid:
     http://www.standupeconomist.com (accessed 13 September 2010).
13   Further details of the Novacem product can be found online here:
     http://www.guardian.co.uk/environment/2008/dec/31/cement-carbon-emissions
     (accessed 13 September 2010).
14   This book was submitted for publication shortly after the 2010 general election, when
     the mental approach to aviation policy of the coalition government was yet to become
     clear.
15   This is the central plot device of the horrifying 1973 distopian science-fiction movie
     Soylent Green.

Bibliography

Ackerman, F. (2009), *Can We Afford the Future? The Economics of a Warming World* (London: Zed).

Ahmed, M. (2009), 'The Next Frontier', *Finance and Development*, 45(3): 8–14.

Albert, M. (2003), *Life After Capitalism* (London: Verso).

Arrow, K., Bolin, B., Costanza, R., Dasgupta, P., Folke, C., Holling, C. S., Bengt-Owe, J., Levin, S., Maler, K.-G., Perrings, C. and Pimentel, D. (1995), 'Economic Growth, Carrying Capacity and the Environment', *Science*, 268: 520–1.

Attwood, K. (2007), 'Tesco funds green consumption studies', *Independent*, 13 September.

Barnes, P. (2001), *Who Owns the Sky? Our Common Assets and the Future of Capitalism* (Washington, DC: Island Press).

Barro, R. J. and Sala-í-Martin, X. (2004), *Economic Growth*, 2nd edn (Cambridge, MA: MIT Press).

Barry, J. (1999), *Rethinking Green Politics: Nature, Virtue and Progress* (London: Sage).

—— (2007), *Environment and Social Theory* (London: Routledge).

Baumert, K. (1998), *Carbon Taxes vs. Emissions Trading: What's the Difference, and Which is Better?* (New York: Global Policy Forum).

Benton, E. (1996) (ed.) *The Greening of Marxism* (London and New York: Guilford Press)

Berglund, C. and Matti, S. (2006), 'Citizen and Consumer : The Dual Role of Individuals in Environmental Policy', *Environmental Politics*, 15(4): 550–71.

Bishop, J. (2008), 'Building Biodiversity Business: Notes from the Cutting Edge', *Sustain*, 30: 10–11.

Bookchin, M. (1971), 'Ecology and Revolutionary Thought', in *Post-Scarcity Anarchism* (San Francisco: Ramparts Press).

—— (1997), 'The New Technology and the Human Scale', reprinted in Biehl, J. (ed.), *The Murray Bookchin Reader* (London: Cassell).

—— (2007), *Social Ecology and Communalism* (Oakland, CA: AK Press).

Boulding, K. E. (1966), 'The Economics of the Coming Spaceship Earth', in Jarrett, H. (ed.), *Environmental Quality in a Growing Economy* (Washington,

DC: Johns Hopkins University Press), http://www.panarchy.org/
boulding/spaceship.1966.html (accessed 13 September 2010).

Cahill, K. (2001), *Who Owns Britain: The Hidden Facts Behind Landownership in the UK and Ireland* (Edinburgh: Canongate).

Carter, M. (2005), 'The MST and Democracy in Brazil'. Working Paper CBS-60-05, Centre for Brazilian Studies. University of Oxford, and MST website, http://www.mstbrazil.org?q=about.

Cato, M. S. (2007), 'Climate Change and the Bioregional Economy', in Cumbers, A. and Whittam, G. (eds), *Reclaiming the Economy: Alternatives to Market Fundamentalism in Scotland and Beyond* (Glasgow: Scottish Left Review Press).

—— (2008), *Green Economics: An Introduction to Theory, Policy and Practice* (London: Earthscan).

—— (2009), 'An International Global Carbon Standard', *Ecopolitics*, Spring/Summer.

—— and Kennett, M. (1999), Green Economics Beyond Supply and Demand to Meeting People's Needs (Aberystwyth, Green Audit).

——, Arthur, L., Keenoy, T. and Smith, R. (2008), 'Entrepreneurial Energy: Associative Entrepreneurship in the Renewable Energy Sector in Wales', *International Journal of Entpreneurial Behaviour and Research*, 14(5): 313–29.

Chambers, N., Simmons, C. and Wackernagel, M. (2000), *Sharing Nature's Interest: Ecological Footprints as an Indicator of Sustainability* (London: Earthscan).

Chan, A. H. (1988), 'Adapting Natural Resources Management to Changing Societal Needs through Evolving Property Rights', *Review of Social Economy*, 46(1): 46–60.

Chertow, M. R. (2001), 'The IPAT Equation and its Variants: Changing Views of Technology and Environmental Impact', *Journal of Industrial Ecology*, 4(4): 13–29.

Costanza, R., Cumberland, J., Daly, H., Goodland, R. and Norgaard, R. (1997), *An Introduction to Ecological Economics* (Boca Raton, FL: St Lucie Press).

Cox, S. J. B. (1985), 'No Tragedy in the Commons', *Environmental Ethics*, Spring.

Cropper, M. L. and Oates W. E. (1992), 'Environmental Economics: A Survey. *Journal of Economic Literature*', 30: 675–740.

Daly, H. E. (1971), *Essays Toward a Steady-state Economy* (Cuernavaca: Centre Intercultural de Documentació).

—— (1977), *Steady-state Economics* (San Francisco: W. H. Freeman).

—— (1999), 'Uneconomic Growth in Theory and in Fact', the first annual FEASTA lecture, Trinity College, Dublin.

—— and Cobb, J. (1989), *For the Common Good* (Boston, MA: Beacon Press).

—— and Townsend, K. N. (1993) (1st edn 1973), *Valuing the Earth: Economic, Ecology, Ethics* (Cambridge, MA: MIT Press).

Davis, L. W. (2007), *Durable Goods and Residential Demand for Energy and Water: Evidence from a Field Trial*, working paper, Department of Economics, University of Michigan.

Douglas, M. (1966), *Purity and Danger: An Analysis of Concepts of Pollution and Taboo* (London: Routledge & Keagan Paul).

Douthwaite, R. (1992), *The Growth Illusion: How Economic Growth Enriched the Few, Impoverished the Many and Endangered the Planet* (Totnes: Green Books).

—— (1996), *Short Circuit: Strengthening Local Economies for Security in an Unstable World* (Totnes: Green Books), http://www.feasta.org/documents/shortcircuit/index.html?sc4/c4.html (accessed 13 September 2010).

—— (1999), *The Ecology of Money* (Totnes: Green Books); also available online at: http://www.feasta.org/documents/moneyecology/contents.htm (accessed 13 September 2010).

Dresner, S. (2002), *The Principles of Sustainability* (London: Earthscan).

Ehrlich, P. *The Population Bomb* (New York: Ballantine).

Ekins, P. (2000), *Economic Growth and Environmental Sustainability: The Prospects for Green Growth* (London: Routledge).

European Commission (2005), *Moving Towards Clean Air for Europe* (Brussels: EU).

European Environment Agency (EEA) (2007), *Air Pollution in Europe 1990–2004* (Copenhagen: EEA).

Fairlie, S. (2009), 'A Short History of Enclosure in Britain', *The Land*, 7, Summer: 16–31.

Fleming, D. (2004), 'The Lean Economy: A Vision of Civility for a World in Trouble', in *Feasta Review 2: Growth: The Celtic Cancer* (Dublin: Feasta), http://www.feasta.org/documents/review2/fleming.htm (accessed 13 September 2010).

Foster, J. B. (2002), *Ecology Against Capitalism* (New York: Monthly Review Press).

Fournier, V. (2008), 'Escaping from the Economy: The Politics of Degrowth', *International Journal of Sociology and Social Policy*, 28(11/12): 528–45.

Frey, B. S. and Jegen, R. (2001), 'Motivation Crowding Theory: A Survey of Empirical Evidence', *Journal of Economic Surveys*, 15(5): 589–611.

Friedman, T. (2005), *The World Is Flat: A Brief History of the Twenty-first Century* (New York: Farrar, Straus & Giroux).

Funtowicz, S. O. and Ravetz, J. R. (1994), 'The Worth of a Songbird: Ecological Economics as a Post-normal Science', *Ecological Economics*, 10(3): 197–207.

Geertz, C. (2005), 'Very Bad News', *New York Review of Books*, 52/2, 24 March.

Goodwin, N. R. (2007), 'Capital', *The Encyclopedia of Earth:* http://www.eo earth.org/article/Capital (accessed 13 September 2010).

Gowdy, J. M. and Hubacek, K. (2000) 'Land, Labour and the Anthropology of Work: Towards Sustainable Livelihoods', *International Journal of Agricultural Resources, Governance and Ecology*, 1: 17–27.

Greenwood, D. (2007), 'The Halfway House: Democracy, Complexity and the Limits to Markets in Green Political Economy', *Environmental Politics*, 16(1): 73–91.

Grossman, G. and Krueger, A. (1991), 'Environmental impacts of a North American free trade agreement', NBER working paper 3914, November (Cambridge, MA: National Bureau of Economics Research).

Hamilton, C. (2006), 'Biodiversity, Biopiracy and Benefits: What Allegations of Biopiracy Tell us About Intellectual Property', *Developing World Bioethics*, 6(3): 158–73.

Hanley, N., Shogren, J. F. and White, B. (2001), *Introduction to Environmental Economics* (Oxford: Oxford University Press)

Hawken, P., Lovins, A. and Lovins, L. H. (1999), *Natural Capitalism: Creating the Next Industrial Revolution* (Snowmass, CO: Rocky Mountain Institute).

Hawthorne, S. (2009), 'The Diversity Matrix: Relationship and Complexity', in Salleh, A. (ed.), *Eco-Sufficiency and Global Justice: Women Write Political Ecology* (London: Pluto).

Heinzerling, L. and Ackerman, F. (2002), *Pricing the Priceless: Cost–Benefit Analysis of Environmental Protection* (Washington, DC: Georgetown University Law Center).

Henderson, H. (1988), *The Politics of the Solar Age: Alternatives to Economics* (first published by Doubleday, NY, 1981, republished by Knowledge Systems, Indianapolis, IN).

—— (2006), 'The politics of money', *The Vermont Commons*, http://www.hazel henderson.com/editorials/politics_of_money.html (accessed 13 September 2010).

Holden, J. and Ehrlich, P. (1974), 'Human Population and the Global Environment', *American Scientist*, 62: 282–92.

Holmgren, D. (2002), *Permaculture: Principles and Pathways Beyond Sustainability* (Hepburn, Victoria: Holmgren Design Solutions).

Homer-Dixon, T. F. (1994), 'Environmental Scarcities and Violent Conflict: Evidence from Cases', *International Security*, 19(1): 5–40.

Hussen, A. M. (2000), *Principles of Environmental Economics: Economics, Ecology and Public Policy* (London: Routledge).

Hutchinson, F., Mellor, M. and Olsen, W. (2002) *The Future of Money: Towards Sustainability and Economic Democracy* (London: Pluto).

Illge, L. and Schwarze, R. (2006), 'A Matter of Opinion: How Ecological and Neoclassical Environmental Economists Think about Sustainability and Economics', German Institute for Economic Research Discussion Paper 619 (Berlin: DIW).

Illich, I. (1974), *Tools for Conviviality* (London: Marion Boyars).

International Monetary Fund (IMF) (2008), 'Globalization: A Brief Overview', http://www.imf.org/external/np/exr/ib/2008/053008.htm (accessed 13 September 2010).

Jackson, T. (2002), *Chasing Progress: Beyond Measuring Economic Growth* (London: new economics foundation).

—— (2009), *Prosperity Without Growth: The Transition to a Sustainable Economy* (London: SDC).

Jevons (1865), *The Coal Question: An Inquiry Concerning the Progress of the Nation, and the Probable Exhaustion of Our Coal Mines* (London: Macmillan & Co.).

Kovel, J. (2002), *The Enemy of Nature: The End of Capitalism or the End of the World* (London, Zed)

Khor, M. (2001a), *Rethinking Globalization: Critical Issues and Policy Choices* (London: Zed).

—— (2001b), *Changing the Rules: An Action Agenda for the South* (Penang: Third World Network).

Kolstad, C. D. (2000), *Environmental Economics* (Oxford: Oxford University Press).

Korten, D. (1995), *When Corporations Rule the World* (West Hartford, CT: Kumarian Press).

Layard, R. (2003), 'Happiness: Has Social Science a Clue?', Lionel Robbins Memorial Lectures, London School of Economics, 3–5 March 2003, http://cep.lse.ac.uk/events/lectures/layard/RL030303.pdf (accessed 13 September 2010)

Leahy, T. (2007), 'Tesco, Carbon and the Consumer', speech to a conference organized by Forum for the Future, London, 18 January.

Leape, J. (2006), 'The London Congestion Charge', *Journal of Economic Perspectives*, 20(4): 157–76.

Lines, T. (2008), *Making Poverty: A History* (London: Zed).

Maddison, A. (2008), 'Historical Statistics for the World Economy: 1–2006, AD' Groningen Growth and Development Centre, University of Groningen, Groningen, The Netherlands, http://www.ggdc.net/maddison/.

Martinez-Alier, J. (2002), *The Environmentalism of the Poor* (Aldershot: Edward Elgar).

——, Munda, G. and O'Neill, J. (2001), 'Theories and Methods in Ecological Economics: A Tentative Classification', in Cleveland, C. J., Stern, D. I. and Costanza, R. (eds), *The Economics of Nature and the Nature of Economics* (Cheltenham: Edward Elgar).

Mayor of London (2002), *50 Years On: The Struggle for Air Quality in London since the Great Smog of December 1952* (London: Greater London Authority).

McGinnis, M. V. (1990), *Bioregionalism* (London: Routledge).

McIntosh, A. (1998), 'Tide must turn for fishing', *Glasgow Herald*, 17 Dec., p. 14.

Meadows, D. H., Meadows, D. L., Randers, J. and Behrends, W. W. (1972), *The Limits to Growth: A Report for the Club of Rome* (London: Earth Island).

Mellor, M. (2006), 'Ecofeminist Political Economy', *International Journal of Green Economics*, (1), 1–2, 139–50.

—— (2010), *The Financial Crisis and the Future of Money: Public, Private or Social?* (London: Pluto).

Mies, M. (1999), 'Women and the World Economy', in Cato, M. S. and Kennett, M. (eds), *Green Economics: Beyond Supply and Demand to Meeting People's Needs* (Aberystwyth: Green Audit).

—— and Shiva, V. (1993), *Ecofeminism* (London: Zed).

Milani, B. (2000), *Designing the Green Economy: The Postindustrial Alternative to Corporate Globalization* (Lanham, MD: Rowman & Littlefield).

Monbiot, G. (2006), Heat: How to Stop the Planet from Burning (Harmondsworth: Penguin). Online: 'on the flight path to global meltdown', *Guardian*, 21 September 2006. (http://www.guardian.co.uk/environment/2006/sep/21/ travelsenvironmentalimpact.ethical living: accessed 15 November 2010).

Morgan, K., Marsden, R. and Murdoch, J. (2006), *Worlds of Food: Place, Power and Provenance in the Food Chain* (New York: Oxford University Press).

Nayyar, D. (1997), *Globalization: The Past in Our Future* (Penang: Third World Network).

Neeson, J. M. (1989), *Commoners: Common Right, Enclosure and Social Change in England, 1700–1820* (Cambridge: Cambridge University Press).

North, P. J. (2009), 'Localisation as a Response to Peak Oil and Climate Change – a Sympathetic Critique', *Geoforum,* 41(4): 585–94.

—— (2010), *Local Money: How to Make it Happen in Your Community* (Totnes: Green Books).

Nsouli, S. M. (2008), 'Ensuring a Sustainable and Inclusive Globalization', a speech to the Universal Postal Union Congress, Geneva, 25 July, http://www.imf.org/external/np/speeches/2008/072508.htm (accessed 13 September 2010)

O'Connor, J. (1988), 'The Second Contradiction of Capitalism', *Capitalism, Nature, Socialism*, 1: 11–39.

—— (1998), *Natural Causes: Essays in Ecological Marxism* (New York: Guilford Press).

Odum, H. T. and Odum, E. C. (2001), *A Prosperous Way Down: Principles and Policies* (Boulder, CO: University of Colorado Press).

Ostrom, E. (1990), *Governing the Commons* (Cambridge: Cambridge University Press).

Pearce, D. W. (1993), *Economic Values and the Natural World* (Cambridge, MA: MIT Press).

—— (1998), *Economics and Environment: Essays on Ecological Economics and Sustainable Development* (Cheltenham: Edward Elgar).

—— and Turner, K. (1989), *Economics of Natural Resources and the Environment* (Englewood Cliffs, NJ: Prentice-Hall).

—— and Barbier, E. B. (2000), *Blueprint for a Sustainable Economy* (London: Earthscan).

Pepper, D. (1993), *Eco-socialism: From Deep Ecology to Social Justice* (London: Routledge).

—— (2010) 'On Contemporary Eco-socialism', in Huan, Q. (ed.), *Eco-socialism as Politics: Rebuilding the Basis of Our Modern Civilisation* (Dordrecht: Springs).

Pigou, A. C. (1920) *The Economics of Welfare* (London: Macmillan).

Polanyi, K. (1945), *Origins of Our Time: The Great Transformation* (London: Gollancz).

Polyzos, S. and Minetos, D. (2007), 'Valuing Environmental Resources in the Context of Flood and Coastal Defence Project Appraisal: A Case-study of Poole Borough Council Seafront in the UK', *Management of Environmental Quality: An International Journal*, 18(6): 684–710.

Porritt, J. (2005), *Capitalism as if the World Matters* (London: Earthscan).

—— (2009), *Living Within our Means: Avoiding the Ultimate Recession* (London: Forum for the Future).

Posner, R. A. (2000), 'Cost–Benefit Analysis: Definition, Justification, and Comment on Conference Papers', *Journal of Legal Studies*, 29(S2): 1153–77.

Robertson, J. (1985), *Future Work: Jobs, Self-Employment and Leisure after the Industrial Age* (London: Gower).

Robertson, J. (1999), 'A Green Taxation and Benefits System', in Cato M. S., and Kennet, M. (eds), *Green Economics: Beyond Supply and Demand to Meeting People's Needs* (Aberystwyth: Green Audit), pp. 65–86.

—— and Huber, J. (2000), *Creating New Money: A Monetary Reform for the Information Age* (London: new economics foundation).

Robinson, P. K. (2009) 'Responsible Retailing: The Reality of Fair and Ethical Trade', Special Issue: Fair Trade, Governance and Social Justice, *Journal of International Development* 21, (7): 1015–26.

Ruitenbeek, J. (1990), 'Evaluating Economic Policies for Promoting Rainforest Conservation in Developing Countries', PhD thesis, London School of Economics.

—— (1992), 'The Rainforest Supply Price: A Tool for Evaluating Rainforest Conservation Expenditures', *Ecological Economics*, 6 (1): 57–78.

Sahlins, M. D. (1972), *Stone Age Economics* (Chicago: Aldine).

Sale, K. (2000), *Dwellers in the Land: The Bioregional Vision* (Athens, GA: University of Georgia Press).

—— (2006a), *After Eden: The Evolution of Human Domination* (Durham, NC: Duke University Press).

—— (2006b), 'Economics of Scale vs. the Scale of Economics: Towards Basic Principles of a Bioregional Economy', *Vermont Commons*, Feb.

Salleh, A. (2009), 'From Eco-Sufficiency to Global Justice', in Salleh, A. (ed.), *Eco-Sufficiency and Global Justice: Women Write Political Ecology* (London: Pluto).

Sarkar, S. (1999), *Eco-socialism or Eco-capitalism: A Critical Analysis of Humanity's Fundamental Choices* (London: Zed).

Schumacher, E. F. (1973), *Small is Beautiful* (London: Abacus).

Shiva, V. (1991), *The Violence of the Green Revolution: Ecological Degradation and Political Conflict* (London: Zed).

—— (n.d.), 'The Neem Tree: A Case History of biopiracy', Third World Network, http://www.twnside.org.sg/title/pir-ch.htm (accessed 13 September 2010).

Simon, J. L. (1981), *The Ultimate Resource* (Princeton, NJ: Princeton University Press).

Simon, J. (1996), *Population Matters: People, Resources, Environment and Immigration* (New Brunswick, NJ: Transaction).

Snyder, G. (1990), *The Practice of the Wild* (Berkeley, CA: Counterpoint).

Solow, Robert M. (1956), 'A Contribution to the Theory of Economic Growth', *Quarterly Journal of Economics*, 70(1): 65–94.

Soper, K. (1996), 'Greening Prometheus: Marxism and Ecology', in Benton, E. (ed.), *The Greening of Marxism* (New York: Guilford Press), 81–99.

—— and Thomas, L. (2006), '"Alternative Hedonism" and the Critique of "Consumerism"', Institute for the Study of European Transformations, Working Paper 31.

Spash, C. (2009), 'Social Ecological Economics: Understanding the Past to See the Future', paper presented to the Association of Heterodox Economists Annual Conference 'Heterodox Economics and Sustainable Development, 20 years on', Kingston University, 9–12 July.

—— (2010), 'The Brave New World of Carbon Trading', *New Political Economy*, 15(2).

Stavins, R. N. (1998), 'What Can We Learn from the Grand Policy Experiment?' Lessons from SO_2 Allowance Trading', *Journal of Economic Perspectives*, 12(3): 69–88.

Stern, N. (2007), *The Economics of Climate Change, The Stern Review* (Cambridge and New York: Cambridge University Press).

Suliman, M. (1999) (ed.), *Ecology, Politics and Violent Conflict* (London: Zed Books).

Sullivan, S. (2008a), 'Markets for Biodiversity and Ecosystems: Reframing Nature for Capitalist Expansion?', paper presented an the IUCN World Parks Congress, 8 October 2008.

—— (2008b), 'Global Enclosures: An Ecosystem at Your Service', *The Land*, Winter, 21–3.

Thomson, D. (2006), 'A Sustainable Future for Small Coastal Fishing Communities', paper presented at A.R. Scammell Academy, Change Islands, Newfoundland at the Change Islands / Simon Fraser University workshop conference Oceans and the Future of Endangered Coastal Communities, 8–10 August.

—— (2008), '*Evidence to the House of Lords' Review of the Common Fisheries Policy*', 19 February.

Thornes, J. E. and Samuel, R. (2007), 'Commodifying the Atmosphere: "Pennies from Heaven"?', *Geografiska Annaler: Series A, Physical Geography*, 89(4): 273–85.

Tietenberg, T. (2006), 'Open Access Resources', *Encyclopedia of the Earth*, http://www.eoearth.org/article/Open_access_resources (accessed 13 September 2010)

UNDP (2007) *Sufficiency Economy and Human Development: Thailand Human Development Report 2007* (Bangkok: UNDP).

UNDP (2007) *Human Development Report 2007/2008* (New York: UNDP)

UNEP (n.d.), *Developing International Payments for Ecosystem Services: Towards a Greener World Economy* (Geneva: UNEP).

Vidal, J. (2008), 'Canadians ponder cost of rush for dirty oil', *Guardian*, 11 July.

Wall, D. (2005). *Babylon and Beyond: the Economics of the Anti-Capitalist, Anti-Globalist and Radical Green Movements* (London, Pluto).

Weizsäcker, E. U. von, Lovins, A. and Lovins, L. H. (1997), *Factor Four: Doubling Wealth – Halving Resource Use: The New Report to the Club of Rome* (London: Earthscan).

Wilkinson, R. and Pickett, K. (2009), *The Spirit Level: Why More Equal Societies Almost Always do Better* (Harmondsworth: Penguin).

Williams, J. B. and McNeill, J. M. (2005), 'The Current Crisis in Neoclassical Economics and the Case for an Economic Analysis based on Sustainable Development', U21 Global Working Paper, http://www.u21global.edu.sg/PartnerAdmin/ViewContent?module = DOCUMENTLIBRARY&oid = 14094 (accessed 13 September 2010).

World Bank (1990), *World Development Report* (Oxford: Oxford University Press).

World Bank (2008), *World Development Indicators 2008* (New York: World Bank).

World Bank (2010), *World Development Report 2010: Development and Climate Change* (New York:World Bank).

Index

Note: Page numbers in *italics* are for tables, those in **bold** are for figures.

CPSIA information can be obtained at www.ICGtesting.com
Printed in the USA
BVOW04s0122020813

327563BV00003B/41/P